"A compelling debut . . . the criminals are quirky rather than terrifying . . . will appeal to readers who enjoyed Susan Orlean's *The Orchid Thief*."
                                        —*Library Journal* (starred review)

"We meet hit men with teddy bears and read about assassination attempts . . . It's intense stuff, and—as crazy as it sounds—it's all real."
                                        —*Pacific Northwest Inlander*

"Spare, exhilarating prose . . . Welch sets up his true-crime tale with a prologue straight out of a paperback thriller."
                                        —*Willamette Week*

"A surprising page-turner . . . Welch's account of the law enforcement officials who hunt these hunters reads like a spy novel, but the vivid imagery is drawn from thorough reporting."   —SmartMoney.com

"*Shell Games* is an eye-opener, exposing a murky world operating just below the surface."   —*The Oregonian*

"Reads like a mix of Sam Spade and *Wild Kingdom*."
                —Jeff Young, host of National Public Radio's "Living on Earth"

"A detective story you shouldn't miss."   —Tom Zoellner

"Forget *CSI*—this is the real deal, tracking down the greediest kinds of criminals as they plunder the planet's future."
                                        —Bill McKibben

ABOUT THE AUTHOR

Photograph by by Meryl Schenker

CRAIG WELCH is the chief environmental writer for the *Seattle Times*. His work has been published in *Smithsonian* magazine, the *Washington Post*, and *Newsweek*. He has won dozens of local, regional, and national journalism awards and has been named the Society of Environmental Journalists' Outstanding Beat Reporter of the Year. In 2007, he completed a fellowship at the Nieman Foundation for Journalism at Harvard University.

SHELL GAMES

A True Story of Cops, Con Men, and the Smuggling of America's Strangest Wildlife

CRAIG WELCH

HARPER PERENNIAL

NEW YORK • LONDON • TORONTO • SYDNEY • NEW DELHI • AUCKLAND

For J & E

HARPER ● PERENNIAL

A hardcover edition of this book was published in 2010 by William Morrow, an imprint of HarperCollins Publishers.

Illustrations that appear throughout the text and the map on p. viii were created by Whitney Stensrud. Grateful acknowledgment is made to the following for the use of photographs: p. 15: top, Joe Volz; bottom, Richard Harrington; p. 19: courtesy of the Washington Department of Fish and Wildlife; p. 59: courtesy of the Joey Skaggs Archives; p. 112: Ed Volz; p. 131: Richard Severtson; p. 179: Jan Jarmon; p. 233: Dean Rutz, *Seattle Times*.

HarperCollins books may be purchased for educational, business, or sales promotional use. For information please write: Special Markets Department, HarperCollins Publishers, 10 East 53rd Street, New York, NY 10022.

FIRST HARPER PERENNIAL EDITION PUBLISHED 2011.

The Library of Congress has catalogued the hardcover edition as follows:

Welch, Craig.
    Shell games : rogues, smugglers, and the hunt for nature's bounty / Craig Welch. — 1st ed.
       p. cm.
    Includes bibliographical references.
    ISBN 978-0-06-153713-4
  1. Poaching—Northwest, Pacific. 2. Smuggling—Northwest, Pacific. 3. Fishes—Effect of poaching on. 4. Fishery management—United States. I. Title.
    SK36.7.W45 2010
    364.16'2863944—dc22

                                                                            2009038980

ISBN 978-0-06-153714-1 (pbk.)

11 12 13 14 15   OV/RRD   10 9 8 7 6 5 4 3 2 1

Contents

AUTHOR'S NOTE

Not long after the publication of *Shell Games*, I caught up with a detective on the other side of the country who had just wrapped up his own wildlife-trafficking investigation. Lieutenant Richard Thomas, with New York's Department of Environmental Conservation, had discovered that leathery creatures by the thousands, mostly snakes and turtles, were being snatched illegally from New York, Pennsylvania, and Ontario. A network of animal traders along the Eastern Seaboard sold them as food or collectors' items. Some rare snakes wound up in Germany, while poachers trucked turtles to a processing plant in Maryland. After the turtles were butchered, their meat was shipped around the world. One smuggler crossed the Canadian border near Niagara Falls with rattlesnakes hidden in pillow cases that were packed in the door panels of his minivan. An entrepreneurial duo from Long Island plucked tens of thousands of turtle eggs from the Pine Barrens and incubated them in sand under heating lamps in their garage. Once hatched, these turtles were eventually shipped to China.

These animals in no way resemble the creatures in *Shell Games*, but aspects of the investigation will sound familiar to readers of this book. Thomas and a colleague spent more than two years undercover, a tactic usually associated with policing cocaine smugglers or tracking the mob. He and other state cops and Federal agents swapped wood turtles in Syracuse, bought copperheads near New Paltz, and confiscated red salamanders poached in Fishkill. Often a great deal of money was at stake: one Long Island turtle-egg hunter told the agents that smuggling earned him $100,000 in a year.

When I finally caught up with Thomas at work on a busy weekend, some of the evidence was crawling at his feet: several of the hundreds of animals he'd seized months earlier still lived in rubber tubs and old fish tanks on his floor. "Let's see . . . I have two box turtles that are about to go to a nature center, six spotted turtles that are supposed to be released this year, and one wood turtle that may have to become our office mascot," Thomas said with a chuckle. "To be honest, I kind of like having them around."

Thomas knew that collectors poached rare reptiles but was surprised to learn, as he told me, that "our little turtles and salamanders are worth quite a bit on the global black market." He was flabbergasted by the sheer breadth of this type of trafficking, as revealed these cases. "It wasn't taking place on the other side of the world, but in our own backyards and in our own communities."

That is precisely the phenomenon I wanted to explore in *Shell Games*. Trafficking in wildlife is a booming international business, but now it is a local business, too. It's no longer just about tigers and elephant ivory or poachers with machetes patrolling African savannahs. Those images are almost comforting in their exoticism. Yes, the United States remains a net consumer of illicit nature—the volume of illegal goods that make it into the country remains larger than what is smuggled out. But creative thieves also find ways to make their fortunes swiping American fish, plants, and animals. The crime caper at the heart of this book involves millions of dollars in stolen shellfish. There are hit men, boat chases, would-be mobsters, mysterious arsons, smuggling kingpins, double-crossing con men, and a police boat called, of all things, *Clamdestine*. (Part of the comedy and tragedy of wildlife crime is that the most bizarre stories also happen to be true.) The events take place in the Pacific Northwest, but that's only because this is the case I chose to follow.

Every corner of the country has something people a world away will pay top dollar to get illegally: orchids and shark fins from Florida, ginseng from Kentucky, glass eels from New England, wild Venus flytraps from the Carolinas, cactus or butterflies from the Southwest, or paddlefish eggs from the Mississippi River. No ocean, desert, forest, or grassland is safe from the poacher's reach. Sometimes these criminals work in the most remote areas of the country. But sometimes you can find them, as I did with *Shell Games*, just by looking outside your door.

Craig Welch
Seattle, 2010

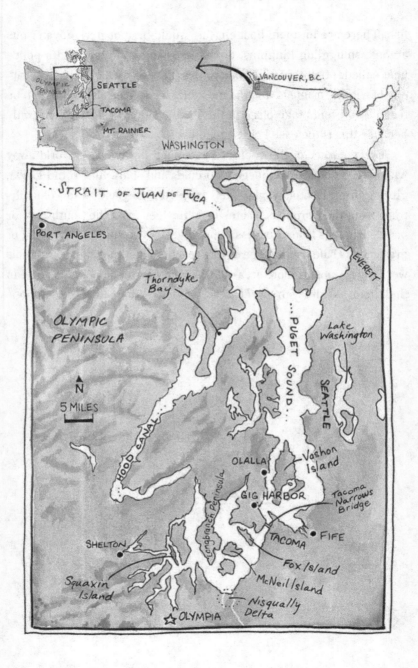

SHELL GAMES

Prologue

THE HUNT

November 13, 2001

The boat didn't look like much. Aluminum with blue trim. A row of smudged cabin windows. A thick center mast crowded with antennas and loudspeakers. Through moonlight and a light rain, Detective Ed Volz could see a curtain of black rubber cloaking half of the vessel's deck like a tent. He couldn't spot the orange glow of a single cigarette and suspected the crew had been ordered not to smoke.

Volz and a partner, Bill Jarmon, were crouched behind Douglas firs and madronas on a wooded bluff overlooking Washington's Puget Sound. They peered down a sandy cliff, Volz through a spotting scope, Jarmon through binoculars, at the boat idling below. Volz heard little other than the wind and the waves. He knew a pair of aging mattresses stuffed in old cloth sleeping bags had been wrapped around an air compressor, muffling its groan. No one who passed by would suspect the crew fed oxygen through a hose to a diver below.

Volz had never been diving. But he knew what could be found in the region's murky underwater world. In the Sound's web of tideflats, channels, marshes, bays, and deltas, life took beguiling forms—particularly in the dimmest depths. Shovel-nosed ratfish patrolled the cobble flats alongside wolf eels with pinched faces that looked chiseled from granite. Anemones glowed in waggling fingers of lavender or in perfect white cauliflower stalks. Ochre sea stars the size of cow heads curled around rocks and mussels, gauging light through the red dots on each arm. Bubble-gum-pink corals camouflaged the porcupine shields of sea urchins.

Rockfish, perch, lingcod, squid. At one time or another, the detectives had found all of these and more in places they did not belong—in nets tied under docks to be retrieved after dark, in aquariums or coolers hidden under tarps in old pickups, on ice in the holding tanks of pirate fishing boats. Thieves hooked, netted, dug, and snatched these creatures and then sold them for food, pets, trophies, even medicine. Some took just a few plants and animals. Others hijacked sea life by the truckload. Volz couldn't recall all the ways he'd seen people steal.

Volz made his living policing the theft of wild things. In twenty-five years with the Washington Department of Fish and Wildlife, he'd chased elk-antler thieves and smugglers exporting bobcat and lynx to collectors. He'd caught poachers who'd hacked off eagle talons for artifact hunters. He carried handcuffs and operated with all the police powers of other lawmen. Here on the western slope of the Cascades, he and his fellow detectives specialized in undercover investigations, mostly involving the region's billion-dollar fishing industry. They'd tracked permits and bank records and trapped Dungeness crab thieves and snared abalone poachers who pried the fist-size mollusks from rocky crevices. But they'd never pursued anyone quite like this captain.

Volz sprawled on a spongy bed of leaves on the bluff and watched and twenty-four-armed sunflower stars, little of consequence hung out above the sandy floor. The real riches were buried below. And while each creature would take a small struggle to retrieve, any discerning thief would gather as many as he could carry. Volz had expected the diver would stay underwater for hours. Then the crew would have hauled up a net carrying a load of seafood large enough to stuff a Volkswagen. Instead, the detectives watched the black shadow peel off his neoprene dry suit.

Within minutes the boat rumbled to life. Hugging the shoreline, it jetted toward Devil's Head point, the last hook on the peninsula before the shoreline looped west and back north.

Volz and Jarmon ran toward the road, clambered into the Expedition, and shot south, trying to keep up. The boat was already out of sight. They drove through the night high above the beach, the truck's headlights off, unable to spy the boat or raise Pudwill on the radio. They rolled the windows down and strained to hear the ship's diesel engine growl through the mist.

The detectives had lost the captain several times before; he used top-of-the-line radar and night-vision gear and moved unpredictably, as if he knew he was being watched. They'd finally caught a break two months earlier. An informant had described crew members on the boat forging documents and illegally hauling in several million dollars' worth of shellfish. If that was true, Volz was watching one of the country's most profitable wildlife smuggling rings—certainly the strangest and most sophisticated the Pacific Northwest had seen in decades. The tipster told them the captain's top shipboard rules: Dump everything if you see anyone approaching; jet away at top speed; and don't stop unless cops thrust guns in your face.

It was good stuff, but nowhere near enough. The detectives had to catch the thieves in the act. Four times in two weeks, they'd tried tailing the boat from land. Each time, they'd followed until 4 A.M. And each fog roll in through the rain and across the black water and the boat below. He could see the faintest glow through the trees across the channel—lights from the prison on the far side of McNeil Island. This investigation kept expanding, drawing in other officers. Tonight, he and Jarmon were joined by a third detective, Charlie Pudwill, who'd made his way to the pea-gravel beach below. Pudwill stood at the water's edge, a quarter mile north. He was closer to the action, but if the boat pulled anchor, he'd have to sprint to his truck and wind up through the woods to the road before he could give chase.

Volz was grateful for the help. It was just after midnight, and the detectives already had tailed the suspect for hours, racing across bridges by land as the boat stole south through Puget Sound. Jarmon was an excellent driver, but he could be a cowboy behind the wheel. He'd cornered the Ford Expedition so fast that Volz had pancaked both hands against the dash trying to keep his head from whipping against the window. The captain had mastered these hidden passages, but Volz and Jarmon found the nearby streets less familiar. They'd bombed in and out of subdivisions, seeking secluded spots with views of the water that were free from neighbors, lights, and dogs. New waterfront bungalows rose among the trees, and construction cones lined streets. Bulldozers had carved a corner off a plot near the road, but no one had started building on this tiny patch overlooking the surf.

Volz adjusted and readjusted his scope. Eventually he saw what could have been a harbor seal bobbing above the water's surface. Then a black-gloved hand emerged, and someone paddled toward the boat. The diver climbed aboard the vessel empty-handed.

Volz was not expecting this. He'd spent hours discussing this very spot with the biologist on his team. Unlike much of Puget Sound, Wyckoff Shoal, just north of Drayton Passage, rarely reached deeper than forty feet. Coarse-grained sand coated the bottom, and tides swept by at two knots. Other than accordion-fanned orange sea pens time, the boat had returned to the Fox Island marina empty. No one unloaded what the tipster said the crew was hunting: the world's largest burrowing clam, known as a geoduck (pronounced "gooey duck"), an obscene-looking giant mollusk that embodied a sea change in wildlife smuggling, a creature with which Volz shared a long and complicated history.

Puget Sound's geoducks had burrowed their way into the Northwest's mythology. Now, thanks to savvy marketing and good fortune and a lust abroad for obscure delicacies, they also had aroused the palates of Asian eaters. Every day, couriers boxed geoducks with gel packs and placed them on jets. Within seventy-two hours they bobbed in restaurant tanks in Beijing or Shanghai or lay in tubs of shaved ice in Tokyo. No matter where the giant clams went they fetched fistfuls of dollars.

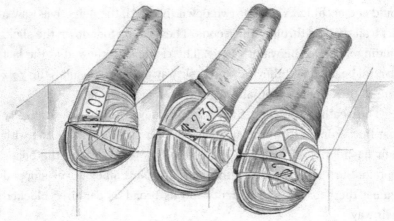

Geoduck, *Panopea generosa*

These bivalves were so valuable that they had been traded for narcotics, and that worth helped create a perfect recipe for crime. With geoduck gathering done underwater and out of sight, corrupt fishermen could take thousands more than the law allowed. Geoducks weren't endangered—they had almost always been plentiful. But Volz knew how fast things in nature could change. Once-abundant sea creatures were declining across the globe, including the Pacific salmon once synonymous with Seattle. The seas were in trouble, thanks in part to overfishing, and Volz had watched geoducks become the region's most lucrative prize. Now these giant clams drew poachers and smugglers and arsonists and hit men, and one audacious thief trailed by a crew of exhausted cops.

Volz and Jarmon drove on through the dark. They pulled over after a few miles, unsure where the boat had gone. They jogged across a clearing to another embankment and peered across the water. They thought they could hear the boat groan in the distance, but the cops could see nothing. A mile or two down the road, the detectives saw a dirt trail that led through the woods. They slid on foot down the slick, winding slope to the water's edge. The rising tide chewed at the last sliver of beach, but if they worked their way to the peninsula's tip, Volz figured they would eventually spy the boat.

The weather worsened. Rain sprayed sideways, mixing with a salty mist that erupted from the surging tide. They could barely see with their flashlights off, but they knew any light would spook the boat's captain. Surf sopped their clothes, and the November waves slopped against the shore. Slippery driftwood as broad as car tires blocked their way.

A mile short of the point, the men stopped beneath a sand-and-clay cliff. Geologists call such cliffs feeder bluffs because their erosion nourishes shorelines with fresh sand. Centuries of tidal pounding had eaten away at the bottom of this wall, undercutting it. Continuing on would be dangerous.

And the men no longer heard the boat. Volz tried raising Pudwill on the radio but got only static. Dark water crashed against the cliff. The beach was gone. The detectives splashed through briny water and foam, which topped their ankles and squished in their shoes. The tide was still rising, and the water this time of year was usually well below fifty degrees. Falling in could lead to hypothermia, unconsciousness, or worse. The detectives had to make a decision.

For much of a decade, Volz had seen his share of mischief. He'd chased lowlifes and hustlers and wannabe gangsters, all of them hunting Puget Sound's inelegant shellfish. Now his ego was engaged; this might be his best shot at this captain. He wanted to get close enough to see the man's face, but the way ran beneath this unstable bank. Volz's watch read 2 A.M., and he and his partner had been working since dawn. Volz was letting the captain mess with his judgment. The men, cold and filthy, stopped. Volz didn't want either of them getting hurt. Reluctantly, the detectives turned and sprinted back toward their truck, clawing through thick brush and stumbling over snags. Blackberry brambles scratched and bloodied their arms and faces. There had been a time for both men when they lived for this part of the job. But now they felt the thwack of every branch.

Caked in mud, they climbed back into the Expedition and stared out the windshield, drained. Rain pounded the hood. Volz felt deflated. He had wasted a lot of time. He brought the radio to his face. Detective Pudwill answered and told Volz he'd been moving around on the beach but hadn't seen anything suspicious in two hours. Volz filled Pudwill in on their fruitless jaunt. Volz and Jarmon would come join him for a moment. Then they'd all head home and get to bed.

Moments later the boat tore back by heading north, its running lights blazing. It shot toward shellfish-rich Wyckoff Shoal. Maybe the captain had gotten careless. Maybe he'd convinced himself he wasn't being followed. The detectives radioed Pudwill one more time. He told them the boat was sputtering into the channel. It cut its engines and its lights as it drifted to the far side. Pudwill radioed Volz that he could see the boat's outline through the predawn haze. It looked as if the captain was ready to drop anchor and get to work.

Jarmon put the truck in gear.

chapter one

~

SNITCHES

Seven years earlier, Ed Volz slumped in a car on the Olympic Peninsula and eyed a Chinese noodle shop from the shadows. It was June 17, 1994, a cool Friday evening, and summer breezes blew in off the gray-green waters of the Strait of Juan de Fuca. Patience didn't come naturally to Volz, but he could summon it when he had to or if something intrigued him, and breaking in a new snitch usually held his attention. Informants are valuable because they're willing to betray. Tracking their movements was one way to avoid getting played.

From his seat, Volz could see the polished glass of the Port Angeles restaurant's front door and his informant's Jeep parked just beyond the glowing streetlights. Volz already knew how this night would go. One officer would record the scene on video from another car. Another would listen from inside the restaurant. And Detective Kevin Harrington would smoke and fidget. Harrington always fidgeted. He hated surveillance.

After ninety minutes Volz saw movement. His informant, Dave Ferguson, popped from the restaurant and jogged down the sidewalk to his maroon Cherokee. Ferguson reached into a cooler packed with dry ice on the backseat and plucked out freezer bags stuffed with shellfish guts. Abalone meat, Volz knew. The palm-size snails live inside barnacle-encrusted shells and are a popular delicacy, especially among Asian foodies. Washington's species, the pinto abalone, had once been so common on the tideflats that residents called retreating currents "abalone tides," but so few remained by the early 1990s that even incidental gathering of them was illegal. Volz had pulled these specimens from a government shellfish laboratory and cleaned and packaged them to look as if they'd come from nearby waters. Now he watched Ferguson stuff the Ziplocs in a brown grocery sack. The informant slipped back into the restaurant, where he resumed talking with his contact, a businessman trying to buy stolen shellfish.

The informant and his contact huddled at the bar's end. Ferguson, stocky and balding, wore jeans and guzzled black coffee. The contact, just shy of 120 pounds and wearing a pinstriped three-piece suit, sipped red wine. He lit the informant's cigarette and told him to stash the shellfish behind the bar. Ferguson's abalone was just a sample. If the contact liked it he would make a regular purchase—at least once a week, he said, ideally in hundred-pound lots. He'd already explained that he would ship the seafood south and have it relabeled as legal California abalone. He'd shown Ferguson photographs and business cards from potential buyers in Vietnam, Singapore, Hong Kong, and China.

The men joked, high-fived, and talked about Asia. Ferguson had been a grunt during Vietnam; the other man had lived on a boat there in the 1970s. They refilled their drinks. The businessman brought out a map, which he and Ferguson studied. If things worked out well, he might take Ferguson to Burma. He would make introductions and show his new supplier the seafood business overseas.

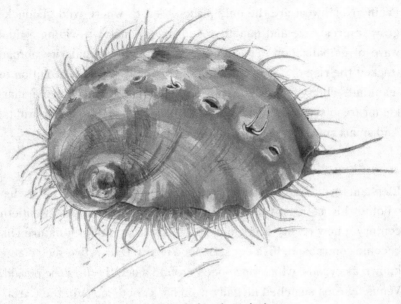

Pinto abalone, *Haliotis kamtschatkana*

Four seats away, another wildlife detective nursed a drink. He was incognito beneath a ball cap, jeans, and a red-white-and-blue slicker but had been watching through a long mirror behind the bar, trying to keep his eavesdropping discreet. Not even Ferguson knew the man was a plant. Volz had put the detective there as backup, since this was Volz's first time working closely with the new informant. Volz wanted to make sure Ferguson relayed events honestly. If he did, maybe next time, the guy could work alone.

The detective at the bar continued to listen. He overheard the businessman tell Ferguson that the two of them would make great partners. "You will learn and I will learn and we will learn together," he said. Then the contact told Ferguson he wanted something else: the big clams with the long necks. The weird ones. The geoducks.

This didn't surprise the cops. The sheltered marine waters of the

Pacific Northwest are the only place on earth where wild geoducks grow in great size and quantity. And the mollusk was riding a tidal wave of globalization. The geoduck's escalating popularity abroad tracked the rise of a new wildlife underground—and an evolution in mankind's ability to exploit nature. In the booming international market for fresh seafood the geoduck had become a path to quick profits. And smart smugglers always followed the money.

～

Elephant tusks, wild furs, alligator skins, and exotic birds. That's what wildlife thieves used to smuggle. But by the close of the twentieth century a new reality was emerging: Almost anything in nature can become contraband. Fish eggs. Baboon noses. Decorative seed plants known as cycads, which have been around since the Jurassic period. Venus flytraps snatched illegally from backwoods bogs in the Carolinas land in stalls at open-air Dutch floral markets. Hundreds of thousands of dollars in illegally caught finger-size glass eels work their way from New England to Japan. Crooks ship stolen monkey blood through Memphis and banned seal oil through Louisville. Illicit fish, plants, and animals of all varieties crisscross the globe to feed black markets.

The transformation came in the last third of the twentieth century, when seismic shifts in the world economy fundamentally altered the nature of global commerce. People used to make purchases through a long chain, buying from local retailers, who bought from wholesalers, who brought goods in from other states or overseas. Then Asian economies ballooned, the Soviet Empire collapsed, and trade barriers fell one after the other. Western-style capitalism washed across continents. Shipping and jet cargo service became routine, along with private package delivery from FedEx and UPS. Industry by industry, technology transformed how everything was bought and sold, increasing the odds that someone bent on selling something could stumble across a buyer somewhere in the world who wanted it. The Internet would only stir things up more, allowing buyers to purchase directly from Thailand on eBay.

All these changes opened avenues for crooks. Suddenly anything from almost anywhere could be purchased outside the law in bulk: pirated Nintendo-game cartridges, fake designer handbags, stolen AK-47s, and plants or animals. Criminal markets and organized corruption went global. Impoverished third-world countries eager for first-world dollars sent poachers to the jungles, and pirates to the seas. By the 1990s, illegal trade accounted for 10 percent of the world economy. The value of black markets as a percentage of the U.S. gross domestic product tripled from 1960 to the mid-1990s. Wildlife trafficking blossomed along with it, becoming, according to the U.S. State Department, one of the largest black markets in the world. The phenomenon would only grow more acute.

Now, every day at U.S. airports and border crossings, wildlife inspectors and customs agents witness inspired displays of duplicity. Bird smugglers stuff stolen live finches in curling irons or squeeze them into socks packed in badminton birdie tubes. Thieves hide tiny golden-throated tropical birds called Cuban grassquits in their underpants or cram banded iguanas from Fiji Island into hollowed-out prosthetic legs. Women smuggle monkeys in their hair. Men hide primates in their pants. Traffickers ferrying bags of sea horses from Mexico cross into the country and immediately mail their booty overseas. Some mark shipping boxes BOOKS to get cheaper rates.

This trade seamlessly intertwines with conventional markets. Boutique fashion outlets from New York to Oregon peddle illegal blue coral jewelry, belts made from alligators killed by poachers, or watch straps carved from the skin of Argentine lizards. Vials of aphrodisiacs ground from deer and tiger penises and smuggled into the country by airline flight attendants sell in herbal markets in major cities. From the tuna in a favorite sushi joint to the gobies in a friend's fish tank, stolen creatures, or goods made from them, sell regularly in shopping malls, pet stores, flower shops, and restaurants, even in supermarkets and warehouses specializing in home furnishings.

The bulk of this trade still comes in from abroad, but the thieving isn't limited to third-world countries. Creatures by the tens of thousands are now being lifted from the forests, deserts, and waterways around America, often not far from major cities such as New York, San Francisco, Miami, Phoenix, and Seattle. Sometimes the criminals prove to be small-time hoods, no different from desperate drug abusers who swipe copper pipes from construction zones. They may take mushrooms from national parks and sell them at farmers' markets or poach deer for meat in a down economy.

But increasingly this trade draws a craftier breed of thief, one literate enough to comprehend obscure laws and wily enough to find creative ways around them. They steal mosses and rattlesnakes and Grand Canyon butterflies, all with an eye toward international markets. These poachers and smugglers piggyback on legitimate markets. They doctor paperwork and shovel loot into shipping containers. They pack stolen goods on jets or drop them with courier services. They trade this contraband like lawful commodities. It is a nearly recession-proof business: In good times, the wealthy demand new delights; when economies tank, enforcement gets curtailed just as people look to nature to stretch their income.

The state and federal investigators who police these traffickers troll rivers, tromp beaches, and hike deserts, forests, and parks. They struggle to halt the siphoning of the country's strangest lifeforms: Pennsylvania turtles, wild ginseng from Appalachia, black bear innards from Virginia, or the seafood residing beneath the waves.

Ed Volz shows off salmon caught
during a fishing trip with his father.

Kevin Harrington strums the
guitar for family and friends.

For most of their careers, Ed Volz and the other Washington detectives had only thought locally. Like narcotics officers nabbing street dealers who sold cocaine through car windows, they didn't care—and didn't need to—about international smuggling. They mostly pinched fish thieves in restaurants and cafés. By law, wild seafood sold commercially has to come from licensed commercial fishermen. Fishermen are supposed to painstakingly record catch volumes and locations, protecting fish from overharvest and consumers from food tainted by pollution or toxins. But cheap, fresh fish caught by local anglers often found its way illegally to local bistros. The detectives camped in dusty boathouses or sprawled in the grass with binoculars, tracking anglers from the shadows. They tailed fishermen to institutions like Seattle's Pike Place Market. They got warrants for business records and backtracked illegal sales, once uncovering a check with "for under-the-table salmon" written in the memo line. Over the years they caught several of Seattle's finest white-linen establishments buying fish illegally. One restaurant raked in fifty thousand dollars a month reselling that illegal catch around town.

But local salmon no longer commanded such high prices. Few of the region's fish did. By 1994, the real money, pound for pound, was often found with marine invertebrates: Dungeness crab, shrimp, crunchy pink sea scallops, pimple-backed cucumbers, spiky sea urchins, and all manner of clams and shellfish. As licensed fishermen sold to an increasingly international clientele, crooks, too, no longer hawked their wares solely to local businesses. Poachers stole by the ton and regularly sold the catch abroad. The detectives were just beginning to grasp the scale of this trafficking and were coming to see that they could use a little help.

~

"He knows the product you are bringing him is illegal?" Volz asked a half hour later.

"Highly illegal," Ferguson said.

"How is he aware of that?"

"I told him," Ferguson said.

It was a few minutes before midnight, and they sat in a dusty, spider-filled back room behind a highway-patrol impound lot a few miles from the Chinese restaurant. Ferguson groaned that he needed to get some sleep. He had plans to get out on the water in the morning. But the detectives wanted to record his version of the night's events while details remained fresh. This was the closest law-enforcement office Volz could commandeer.

The Olympic Peninsula juts like a mitten thumb from Washington's mainland, with the city of Port Angeles centered on the thumb tip. The city is the timber-rich peninsula's last gateway: a direct ferry trip across the choppy strait from Victoria, a short drive from the hiking trails and rain forests and granite ridges of Olympic National Park, and a longer journey by road and boat from Seattle. Both a tourist oasis and a working waterfront, it has gift shops and restaurants, a Coast Guard station and a lumber mill. In 1994, it remained small enough that shopkeepers often knew their customers and busy enough that strangers usually went unnoticed. Volz and Harrington counted on that.

The men huddled in the poorly lit building. The buyer had been so skittish, Ferguson told them, that once they strayed beyond mere talk, he insisted that Ferguson speak in code. They agreed that from now on they would call the shellfish "bolognas"—just in case. The buyer kept telling Ferguson to keep his voice down. But he agreed to meet later and buy more stolen shellfish.

Volz was as interested in Ferguson as he was in the buyer. So far the detective had to admit he was impressed; Ferguson had pulled it off. His description matched perfectly what the cop had heard from the bar. The detectives were encouraged by the performance. Despite Volz's skepticism, Ferguson might prove valuable. The detectives knew they could use the assistance.

An anonymous tip had led Volz to Ferguson, a poacher who'd slipped by unnoticed for years. In March 1994, a caller alerted police to a suspicious diving boat puttering around a tranquil bay along the strait. Police watched from afar, spying on the twenty-six-foot boat until it docked in Port Angeles. The cops saw a squat, balding man with a mustache as thick and brown as a cigar hand a sack to a friend on the boat. The pair unloaded their gear and carried off the sack and a tub covered by a white cloth. Police confronted them and found 188 stolen abalone—not a huge case, but worth prosecuting.

Notified of Ferguson's arrest, detectives Volz and Harrington headed west from Seattle the next day. They crossed Puget Sound by ferry and motored up Highway 101, which circles the rim of the Olympic Peninsula and parallels the Pacific Ocean all the way to Southern California.

The two detectives had policed small-scale fish thieves together for years, but Volz and Harrington could not have been more different. Volz had grown up in Seattle's suburbs and liked to hunt and fish. He was a barrel-chested block with a round face and coarse dark hair that gave him the look of a shorter, friendlier Charles Bronson. He had been an animal cop since 1976, and in the fleece-chic Northwest, he concealed his .45-caliber service pistol beneath windbreakers and patterned shirts. He had studied fish biology at the University of Washington, and for a cop had a sophisticated understanding of marine life. He was serious, thorough, and rarely cut corners. But he had a lip and often squabbled with his bosses. Even the drafting of a search warrant could get him grousing about the job's minutiae.

Kevin Harrington (*standing*) and Ed Volz fish while working
undercover on an unlicensed charter boat in the early 1990s.

Harrington came from Michigan and considered himself a softy.
He laughed often and cried openly at movies, rating them for his wife
by degrees of "weepiness." Once, working undercover alongside fish
thieves on an illegal charter boat, Harrington caught some of the
biggest fish of his life. In the parking lot he moved to tip one of the
poachers twenty bucks until his flabbergasted partner asked: What
are you doing? Harrington shrugged: Just being polite. Harrington
chain-smoked, played guitar, and hated wearing his gun. His superi-

ors eventually would dress him down because he so often left his side-arm packed away in his trunk.

The two cops irritated each other but worked as a team. Volz did not mind stakeouts but hated sifting through records. Harrington actually enjoyed forensic work; he had the patience Volz lacked and liked smoking while deciphering documents and believed that even the smartest criminals left a paper trail. When interviewing suspects, the two cops found a routine. Volz enjoyed needling hardheads. Harrington established rapport. Volz called him "the counselor" because everyone confided in him. Their tag-team style grew so instinctive that a change of expression by one could inspire the other to shift tactics. They usually found ways to get what they wanted.

Dave Ferguson surprised them both. He had told the arresting officer he would admit guilt and help wildlife cops catch the people who bought his contraband. All he asked was that they free his boating partner, who was unaware that Ferguson had been poaching. Volz and Harrington plucked Ferguson from jail and took him to a nearby office. Once assured their interest was solely in him, Ferguson answered every question without a fight.

Ferguson admitted that he regularly stole abalone, which in Puget Sound mostly live between the surface and waters thirty-five feet deep. Ferguson said he frequently covered up his poaching by scuba diving legally for sea cucumbers. His latest haul had been destined for a Chinese eatery in Port Angeles, but he also sold to a cantina in the port city of Everett, north of Seattle. Workers there kept scales inside the back door and weighed illegal products, paying cash from the till. Ferguson knew some of them resold the shellfish overseas.

Ferguson had a record, but he also had kids. His girlfriend was pregnant, and his bank account was empty. He faced criminal charges and the loss of his fishing license and livelihood. The detectives saw an opportunity. A thief desperate enough to be honest might be desperate enough to be useful. And the loyalty Ferguson had shown his friend gave Volz hope. Unlike the hard cases usually seen on television, the criminals Volz knew would sell out anyone to escape trouble. Ferguson possessed some sort of moral compass, even though he was no minor crook. He had shot at police in California, served as vice president of a prison motorcycle gang, and had faced federal charges for running guns.

Ferguson could be just what they needed. He had confessed to stealing enough abalone over the years to buy his Jeep and pay off his boat, defiantly named the *Abalone Made*. Biologists later calculated that Ferguson alone may have taken as much as all recreational fishermen combined had in previous years. They suspected he may have swiped twenty-five thousand abalone—nearly all that were left in the waters near Port Angeles.

And no one believed he was Puget Sound's only major shellfish thief. The cops suspected that there might even be a shellfish kingpin, a seafood broker to whom poachers funneled their stolen loot. The detectives were on a mission to find out. After administering a polygraph, they offered Ferguson work as a paid informant. Given his reputation, Ferguson offered a unique chance to assist them. No one who knew him would suspect he'd cooperate with the police.

A month after his discussion about geoducks and "bolognas," Ferguson returned to the Chinese restaurant, this time with an undercover female patrol officer posing as his wife-to-be. The informant's discussions with the businessman had continued, but Ferguson and the detectives had begun shifting gears. They'd started gathering intelligence on more established smugglers and had come to realize that this contact was just a minor player. They wanted one more meeting to be sure.

The three spoke over dinner about a party the buyer was throwing for friends and discussed once again how best to ship geoducks overseas. Each time Ferguson suggested they slip out to his Jeep to get the illegal shellfish, the buyer waved him off, suddenly uncomfortable cutting deals in public. Finally the businessman offered an alternative. "We will go up to my house," he said.

They piled into Ferguson's Jeep and drove to a split-level home in the hills near Port Angeles. They entered through a sliding door and headed for the basement, where Ferguson and the officer handed over more seafood. When the buyer opened his freezer, Ferguson looked inside. There, still in Ziplocs, he saw the abalone he'd sold in June. The buyer hadn't eaten or resold any of it.

Volz faced a decision. The businessman may have had international ambitions, but clearly he was still small-time. Volz could pursue minor charges against this buyer, but that would mean Ferguson would have to testify. Volz hated the thought of exposing his informant. Ferguson had been making contacts every week with fishermen and seafood brokers throughout the Northwest. He had gotten so good at digging up information that Volz saw a chance to make cases that were far more sophisticated.

Volz made a pitch that his agency should go big. If his bosses spent a few hundred dollars overhauling Ferguson's boat, they could send him to work undercover as a geoduck fisherman. Thanks to Ferguson, Volz wrote in a report to his superiors, "the opportunity exists to have a huge impact in the illegal commercial shellfish industry." The informant could give detectives an inside line on illicit trade in the world's largest burrowing clam. Volz's superiors quickly agreed. They bought equipment and paid for Ferguson to get licensed as a geoduck fisherman. Right away, the informant took work with a Canadian shellfish company. He hit the water on September 10, 1994.

The call about the fire came the next day.

Ferguson's girlfriend screamed at Volz through the phone. There had been an explosion on the *Abalone Made*. Ferguson was being ferried by helicopter to Harborview Medical Center in Seattle. He'd been in the wheelhouse gassing his boat at a fuel bay on the Olympic Peninsula. When he turned the ignition the flash blew out the boat's windows and hurled Ferguson through the air. Clothes black and tattered, he shuffled zombielike toward the street. Firefighters would describe a charred and hobbling monster. Second- and third-degree burns covered the bottom half of his body.

Doctors found the damage confined mostly to his legs. The informant needed skin grafts and faced weeks of painful recovery. At Ferguson's insistence, Volz and Harrington visited the burn unit. They fussed over the wounded criminal for weeks like concerned parents. He thanked them profusely, sometimes blubbering like a child. The detectives took shifts at the hospital as Ferguson underwent surgery.

Volz didn't know what to think. Was the fire an accident or sabotage? Who knew Ferguson was a snitch? He learned through back channels that fire investigators had mixed theories. No one could dismiss the possibility that Ferguson had been targeted. Volz's superiors had questions, too, but Volz insisted the agency use discretion. If wildlife detectives snooped around the fire, it would only raise questions that could put Ferguson in more danger.

Three weeks later, when Ferguson left the hospital with a hundred square inches of fresh skin attached to his legs, investigators didn't know much more about what had caused the explosion. Loaded on painkillers, Ferguson walked the remains of his boat with a deputy sheriff. They found a fuel-tank vent hose lying in the bilge; it should have been clamped to a fitting in the hull that vented fuel over the side. Explosive vapors had leaked into the engine room and sparked when Ferguson turned the ignition. Had it jostled loose or been intentionally dismantled? "There is no explanation as to how the vent tube ended up unhooked," the deputy wrote in his report. The fire's origins would never be resolved.

Ferguson emerged from the incident grateful for the attention Volz had paid during his hospital stay and begged Volz to let him continue as an informant. He said it gave him a sense of mission. That fall and winter, Ferguson went back undercover, keeping tabs on geoduck fishermen and helping set up small buys of illegal seafood.

There is no accounting for the strange relationships that develop between cops and informants. Ferguson treated Volz like a sibling. He shared petty grievances. He fretted about a friend who landed in the hospital with the bends. He complained about his girlfriend. He whined about the cost of a ruined boat propeller. He confided to Volz that the guys he knew in prison usually cheered for the cops on TV.

Volz listened dutifully and wrote it all down. He could be brusque and combative when he thought Ferguson was out of line, but he developed a grudging respect for the informant. Not that Volz actually trusted him. He made the informant contact him daily. Ferguson had to keep meticulous notes. Not only could he not make illicit transactions without approval, but he also could make few decisions entirely on his own. Every move first had to go through his handler. Ferguson didn't mind and kept delivering through the end of the year, offering new tidbits that led detectives deeper into the geoduck trade.

On a wet January morning six months after his surveillance at the Chinese restaurant, Volz slumped in a vehicle once again, this time behind the darkened windows of an SUV. It was 1995, and Ferguson's work had led Volz to a highway park and ride south of Seattle where he planned to videotape an illegal sale from a safe distance. But a red and silver Honda CRX screeched into the adjacent spot, and Volz found his suspect working two feet away.

Volz tried not to laugh. The driver struggled to jam several fifty-five-gallon cans of stolen geoducks into a car barely large enough to pack a bag of groceries. Wet shellfish dribbled onto the backseat. Then the driver peeled out with the hatchback propped open. Other detectives fell in behind in unmarked cars.

The man made a stop, transferring his cargo to a white van, which then hit Interstate 5 and headed south. By the time Volz caught up with the crew following the van, it was crossing multiple lanes of traffic and slipping quickly toward exits only to immediately pull back onto the highway. Volz had expected a modest sale of stolen shellfish, but this guy moved like a professional, cutting and weaving as if trying to lose a tail.

Volz and the other fish cops followed for three hours from Washington into Oregon, fearing with each mile that they would run out of gas. When he crossed the border with stolen goods, the driver had committed a federal crime. In Portland, the men watched the driver stop and transfer his load into a midsize seafood warehouse. That suggested to Volz that they had stumbled onto a network. The detectives backed off, not sure of the scope. They already had been communicating with the Feds. This new development confirmed that they were on to something new, but figuring out exactly what might take some time.

The detectives regrouped. Weeks later they confronted the van's driver, who confessed that he'd bought stolen clams for years. He said he shipped them to the Oregon warehouse, where buyers then shipped them to Brooklyn. This driver told detectives he was not alone. Thousands upon thousands of illegal shellfish—maybe more—were regularly changing hands on the water, then crossing state lines and moving overseas.

There was no telling how big this mess could get, but Ferguson wouldn't stick around to find out. One day that winter, in 1995, a fellow fisherman accused him of being an informant and rammed his boat. Days later another fisherman accosted him. Ferguson that spring finally called Volz in a panic. Volz always knew this day would come. Anxiety from the explosion was resurfacing. Ferguson made clear that he wanted out. Volz reassured his informant, but he could tell it didn't take. Working for the cops took time yet paid only a meager stipend, and Ferguson no longer had the same zeal. He wanted to move on and start a new life, and Volz knew he couldn't really demand much more. Within months, Ferguson quit. He packed up and moved to Alaska. Volz would never see him again.

Volz considered the implications for his investigation. Ferguson had been a great inside player—an outlaw with a reputation who got them close to dirty fishermen. Most of the leads they now had originated with Ferguson. The man had charm, chutzpah, and intelligence. Informants like that didn't come along often, and Volz couldn't imagine where he would ever find another.

chapter two

LARGER THAN LIFE

Inside a restaurant on a sunny day five months later, several men packed a table and ate seafood. They were big guys and they anchored the table like sturdy pilings. Doug Tobin, as always, did much of the talking. Tobin had telephoned diving instructor Dennis Lucia asking for help becoming a commercial diver. Lucia had agreed that they should meet here, at Pearls by the Sea, in the southern Puget Sound village of Purdy.

It was a hot August afternoon, between the lunch and dinner rush, and the manager fluttered about while they talked. He readied the pie case and milkshake machine for the coming crush of summer cabin renters. In a few hours the dinner patrons would crowd in, demanding salmon, fried prawns, steak, and crab. For the moment, though, the men had the dining room to themselves.

Doug told Lucia that he and his brother John were Native Americans from Puget Sound's Squaxin Island Tribe. They made their money catching fish, mostly salmon. The brothers netted chinook in

August, chased coho into October, and hunted chum through early winter. Between seasons they often logged in Alaska. Doug had operated heavy equipment at gold mines and construction sites, but mostly the brothers stitched together lives from the sea, seining for perch, hauling halibut and lingcod, or digging manila clams from the beach. Now they wanted to dive for shellfish.

The men talked and ate. The restaurant overlooked the water, and through rows of windows they could watch tides lap against Purdy Spit, a mile-long sandbar that corralled a huge brackish backwater lagoon. Gray whales, frequent visitors to Puget Sound, sometimes strayed into this lagoon and swam in circles for days before finding their way out. Lucia noticed that Tobin spoke in a circuitous stream and got lost in tangents before getting back to his point, but he usually managed to get his audience to hang with him. Three words rattled in Lucia's head as he sized up his new client—*larger than life*.

Doug Tobin had always been big, but at forty-three he was as dense and meaty as an aging pro wrestler. He kept his black and silver mane long, and it curled at times into ringlets that piled up on one another. He could have passed for Louis the XIV, if the Sun King had favored flannel shirts open to his ribs and had worn a whale-tooth necklace.

Tobin possessed a wrestler's showmanship, too. When he told stories he performed them with arched eyebrows, elbow squeezes, winks, and one-liners. He dismissed an ex-business partner with a wave: "No whale in the ocean has a bigger blowhole than that guy." Emphasizing a point, he frequently cupped a hand to his ear: "What's that you say?" He would then proceed to answer his own question. He followed the news and spoke perceptively about world affairs but had quit school early and was prone to malapropisms. He frequently exploded with off-color comic rants and introduced his stories with such fanfare that few knew which were true: "What I'm going to tell you will blow you out of this room."

Doug and John had been raised in logging camps around hard

men. They drank coffee in grade school and slept through recess, tired from cutting trees after class. When fishing icy waters, Doug had been known to urinate on his hands for warmth. Nicks and cuts scarred his big palms after years of grabbing fish by their razor-toothed mouths. The boys had grown up fast and with exacting standards. When school shut down the week of JFK's assassination, their father had the brothers net salmon for money. Another time, Doug's father professed his annoyance that Doug couldn't yet fix the car's brakes. Doug was ten.

Doug's talent and ego and the demands placed on him as a kid propelled him toward a certain perfectionism. Fascinated by Native American wood carvings, Doug pared Salish masks and totem poles from cedar planks and cut intricate pieces that landed in museums and galleries. He was one of the Northwest's best fishermen. He worked hard enough that he and a friend once slept in puddles on the floor of a cramped topless skiff during a hard rain while gathering salmon. It was a miserable night for the friend, who woke and found Doug happily hauling in a full net.

Both brothers could be intense—John spoke rarely and smiled less—but Doug was a charmer, a prankster, a benefactor, and a bully. He bought diapers for needy neighbors and Easter dresses for a struggling single-mother's daughters, and when his neighbor's business had a bad year, Doug quietly filled the family's freezer. He taught people to fish, sponsored youth teams, worked with troubled tribal teens, and was quick to loan money to those in need. Yet his teasing was merciless and sometimes bordered on cruel. As a kid he talked friend Steve Sigo into getting plowed on hard liquor and then for kicks told Sigo's father what his son had done. Another time, Doug, returning from a hunting trip, found Sigo digging clams on a beach. He fired several shots in Sigo's direction, apparently just to watch his friend squirm. Though Sigo knew Doug wouldn't actually shoot him, Doug was the only kid he ever ran from. As an adult Doug could be physically menacing and was sent to prison twice for violent crimes, but he was often disarmingly carefree and magnetic. He could be arrogant or self-effacing, at times in the same sentence, his broad mouth hinting at laughter, whether or not he was amused. An entourage of buddies trailed him like a prophet. A handful of friends called him Elvis.

Listening to him in the restaurant, Lucia had to admit it: He liked Doug immediately. The man was funny and good-natured, a natural salesman and entrepreneur. Too many would-be divers approached the trade casually, but the Tobin brothers took the idea of dive training seriously, as they did anything that involved fishing. The men wanted to learn fast, but Lucia could tell they wanted to learn right. Puget Sound was in the midst of a major shellfish boom. They wanted to get trained because they thought they were missing out. And they were.

Nearly a century and a half earlier, Washington's first territorial governor had swindled Native tribes, giving them cash and taking land and pushing them onto reservations. He signed treaties that ensured Indians retained rights to gather fish "in common with" non-Indians— a phrase that would be fought over for nearly twelve decades. Fishing in the Pacific Northwest would become a path to riches, but whites came to control the entire industry. In 1974 a federal judge changed everything. He reaffirmed treaties that gave Puget Sound tribes rights to half the catch of fish. Less than a year before Tobin met Lucia, another judge extended the ruling to include marine invertebrates, spineless creatures like razor clams and butter clams, other shellfish and crab. Tribes had only to follow particular rules.

The Tobin brothers saw the exploding demand for geoducks. Like others, they wanted to capitalize on the rush and join Puget Sound's growing armada of clam divers. They were eager to squeeze into wet suits and head beneath the waves.

After the world's largest ocean pushes past the granite peaks and ancient conifers that line the Strait of Juan de Fuca, it plunges on toward an archipelago of nearly eight hundred small islands, formed by the tips of underwater mountains. Half of the San Juan Islands are so tiny they disappear under swollen tides. The visible ones poke through mounded and lumpy, like the rough backs of giant half-submerged crocodiles. South of these islands, the sea curls into a comma, and the blue-gray ocean becomes Puget Sound.

The Sound stretches north to south more than one hundred miles. It swirls along crooked shorelines and around serrated peninsulas, pushes into fingery inlets and fills a long channel called Hood Canal. This creates so many miles of squiggly coastline that, if unfurled, it would stretch from British Columbia to the tip of the Baja Peninsula. Pacific seawater splashes against these shores and washes with fresh snowmelt pouring in from ten thousand creeks and rivers. That water ferries plants and insects from disparate terrains—the snowfields of an active volcano, the fertile mossy duff of rain forests, remnant bunchgrass prairies. The food and snowmelt churn with icy jets of sea-water, which rise from depths of up to nine hundred feet. It's a cocktail that nourishes one of the most ecologically productive marine systems in the world. "Nothing can exceed the beauty of these waters," U.S. Naval Captain Charles Wilkes wrote upon entering Puget Sound for the first time in 1841. "I venture nothing in saying there is no country in the world that possesses waters equal to these."

The Sound is awash in the beautiful and the bizarre. Porpoises and killer whales skip across the surface. Crystal jellyfish pulsate in the depths, aglow with ethereal light. There are pink lumpsucker fish shaped like Ping Pong balls, and weird noise-making swimmers: grunting sculpins, croakers that drum and groan, and a toadfish called the plainfin midshipman, which coaxes females into depositing eggs by humming.

Even in this world one peculiar organism stands out: *Panopea generosa*, the geoduck. Glistening and pale, or coarse and leathery as a bicycle tire, the geoduck's ribbed neck, or siphon, can stretch the length of a bowling pin or recoil to a fat and wrinkled nub. The clam is immense and, to many, vile, resembling an elephant's trunk, an aardvark's snout, or, most obviously, the reproductive organ of a Clydesdale stud. Few who've seen a geoduck forget the experience.

The geoduck is a Pacific Northwest celebrity and a lurid punch line—oversize, ugly, and still somehow charming. One boutique seafood wholesaler pitches geoducks as an erotic gift "sure to lead quickly from the dinner table to the bedroom." A cable television show host tweaked Mae West's famous wisecrack: "Is that a geoduck in your pocket or are you just happy sashimi?" Holding a geoduck on a visit to Seattle in the 1980s, the *Today* show's Jane Pauley quipped, "God does have a sense of humor."

Such attention can seem excessive for a bivalve that wiles away decades buried in muck. The geoduck's name comes from the Salish

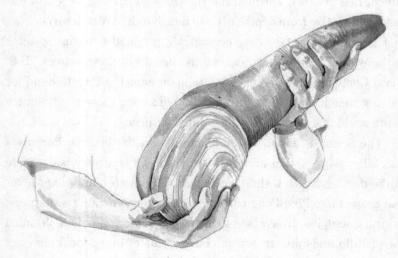

Geoduck, *Panopea generosa*

Indian phrase *gweduc,* meaning "dig deep," which is what young geo-ducks do. A baby geoduck uses a tiny muscular foot to tunnel slowly into the seafloor. Over time, the geoduck gets bigger and burrows deeper. Eventually its shell ends up cocooned several feet down with only its neck poking through the mud and into the water. Here the clam will remain, fixed in its burrow until death.

Geoducks can grow to fourteen pounds and live more than 150 years. (The oldest geoduck lived to be 168. Another clam species, *Arctica islandica,* the ocean quahog, is the planet's longest-living animal. One found off Iceland lived four hundred years.) Geoducks produce growth rings on their shells much like rings on trees, and research-ers measure their width to track global climate changes. Puget Sound geoducks can so accurately reflect severe weather across the globe that a clam plucked from the Northwest can offer a fossil record of a poor monsoon season that sparked famine in India a century earlier.

Smaller related clams exist in Argentina, Japan, and New Zealand, but Puget Sound is the geoduck capital of the world. Rows of geoduck siphons protrude in clusters from the seafloor like old-growth forests of the sea. As with Douglas fir trees or king salmon, these mollusks are a symbol of Northwest pride, an elaborate caricature of the region's abundance. Evergreen State College in the capital city of Olympia, a progressive alternative school that eschews grades and collegiate con-ventions, claims the necky clam as its mascot. At basketball games, stu-dents chant, "Go, geoducks go / Through the mud and the sand, let's go." In a speech during the faculty's founding retreat in 1971, botany instructor Al Wiedemann lauded the geoduck's craning neck. "Now that's real flexibility," Wiedemann said to rolling giggles. "And what is Evergreen but flexible?" In *Three Feet Under,* documentary filmmaker and Seattle native Justin Bookey's love letter to the clam, Wiedemann, in a sweater and horn-rimmed glasses, leans against a podium and describes how the geoduck embodies the spirit of Evergreen's mission. "His most compelling quality is that he's totally nonaggressive," Wiedemann says. Hence the laid-back school's motto: *Omnia Extares,* "Let it all hang out."

Geoducks can be dug from the mudflats during the lowest tides of the year, and in early summer one waterfront bar, The Geoduck Tavern, hosts a contest: The patron plucking the biggest clam drinks for free. But most of the Northwest's millions of geoducks are found in deep water, and fishermen with diving gear have legally collected them for sale as seafood since 1970. Because geoducks live so long, it's difficult for scientists to measure the long-term impacts of fishing, so commercial geoduck diving is tightly regulated, with strict limits set on when and where divers can take clams. Laws limit divers to daylight hours so official government monitors can watch the harvest and make sure no diver takes more than the quota.

Through the pinking gray of dawn these divers come each day in specialized boats with the cabins pushed forward to clear space for diesel generators and air compressors. With names like *Gold Rush* and *Rawhide,* the thirty-two-foot trawlers and forty-foot charter boats slice through rolling chop. The fishermen, in jeans and tattered sweats, cotton hoodies and fleece vests, anchor their boats off designated spots in fir-lined peninsulas and coves. Amid lines, hoses, and chest-high stacks of plastic milk crates, divers clip seventy-pound weights to their waists or ankles. They wear heavy boots instead of flippers, and when they slip into the wash they breathe through an umbilical hose attached to an air supply aboard the boat. The law restricts them to waters less than seventy feet deep; going deeper would increase the odds that they would need to recover in decompression chambers.

An electronic intercom links divers to the deck, where their tinny grunting can be heard amid the hiss and pop of loudspeakers. Attentive captains can determine a diver's well-being by listening to each breath. Down below, the divers crawl over the sand armed with a water spray gun called a "stinger." They look for the tip of a geoduck siphon, a tiny figure eight poking through the muck like binocular lenses. Then the diver grabs this flesh and turns the water gun on the sand to clear it away. The diver frees the clam from the mud and packs the catch in a bag the size of a one-man hot-air balloon, which he sends to the boat on a line hauled by a winch. By the time the divers pop above the surf, the crews onboard are already placing the geoducks in plastic crates, which they heave on top of one another until the clams pile up like glistening stacks of hundred-dollar bills.

❦

When it came to training divers, Dennis Lucia was an obvious choice. He had worked below the waves since the 1970s, removing batteries and hazardous material from the Gulf of Alaska for the Coast Guard. He worked the Prince William Sound cleanup when the *Exxon Valdez* dumped eleven million gallons of crude in 1989. By the early 1990s, would-be clammers jammed his answering machine, offering bonuses if he moved them ahead in line or would squeeze months of training into a weekend. By the time Lucia heard from Doug Tobin, he already had scheduled a class for Squaxin Island Indians. He asked tribal leaders if he could add the brothers. The tribe said no. The Tobin boys, he was told, did not always play well with others.

Lucia, unperturbed, agreed to give the brothers a special class. They met at John's house across the spit. The Tobin meticulousness was everywhere in the home: the grass trimmed and edged, the outbuildings freshly painted and scrubbed. In a spotless kitchen, Lucia went over written material—pressure equalization, nitrogen narcosis. John asked questions. Doug barely paid attention. Still, when Lucia started grilling the brothers, Doug recalled the essence of every lesson.

They rocketed off on a sunny morning, puttering into the sheltered waters of an isolated shark-fin-shaped bay. John dived first and

struggled to orient himself. When his turn came, Doug took obvious pleasure in being underwater. Geoduck fishing requires patience
and a supple grasp of body mechanics, the kind that astronauts
need when grappling with lug nuts. Fishermen wrestle constantly
for leverage as they drag hundreds of feet of heavy line in a nearly
weightless environment. When currents grow swift, the divers strain
to stay put, jamming their spray nozzles, their stingers, in the ground
for balance like mountaineers leaning on ice axes. Inattentive divers
get tangled in their lines or wrapped into balls around boat anchors.
Some complain of being lifted off the bottom by tidal forces that feel
like hurricanes.

That environment can kill, and often has. In 1988 a seafood diver's
weight belt came undone below, sending him so quickly to the surface
that he fell unconscious, rolled onto his face, and drowned. The next
year, a diver surfaced with no mask, one fin, and blood rushing from
his ears, nose, and mouth. He died within minutes. Another diver's
weight belt and water line cinched around his air hose, suffocating
him as effectively as a noose. Whales crash against geoduck boats and
pin divers on the bottom. A gray whale once nosed through the silt
and actually struck geoduck diver Mark Mikkelsen. Earlier in the day
he had seen the whale grinding its nose in the mud and gyrating and
twisting its body. Perhaps attracted to the vibration of water through
Mikkelsen's hose, the whale later plopped down on Mikkelsen's lines,
trapping him in place. A few seconds later, the creature lost interest.
Thinking it was gone, Mikkelsen resumed digging. Then the whale
smacked his back, flattening him as if he'd been slammed by a two-by-
four. After righting himself Mikkelsen watched the barnacled snout
nose back toward him through the silt. He could have reached out and
swatted it. The whale hovered momentarily and was gone.

Doug did not seem unnerved by the dangers. Instead he focused
on the world opening before him. Geoduck divers saw sights few oth-

ers did: rare territorial sixgill sharks, some reaching sixteen feet, and great Pacific octopuses with flaming red bodies. Divers came whisker-to-whisker with barking thousand-pound Steller's sea lions, which could appear, in confrontation, like underwater grizzlies. While geoduck fishermen worked, spiny dogfish, small sharks with mildly toxic dorsal fins, swarmed their mesh bags of geoducks like packs of hungry Rottweilers, the fierce nippers bumping divers in the rear.

After their training, Doug and John Tobin motored their boats in and out of coves while they gnawed on cigars, muscles rippling under their tank tops. A fishing buddy helped set up John's boat while Doug found his own partner and tried to gain a competitive edge. He planned to get a seafood broker's license and buy from Indian divers. He suspected they would offer him discounts if he paid in cash.

Almost immediately, Doug said, he saw corruption. Everywhere he looked someone ran a scam. Scuba shops sold divers ancient weight belts, cheap dive computers, and tattered dry suits. "Everybody blew smoke up our asses," Doug later said. "It started with the dive shops. They'd have you getting into this big old huge bullshit suit that made you look like Winnie-the-Pooh. Nobody bothered saying, 'If you want a real suit, get ready to drop three thousand dollars for crushed neoprene titanium with Kevlar knees and lugger boots.' We weren't talking about weekend-warrior diver bullshit. We were talking about commercial harvest!"

Doug settled on gear and focused on finding clams. Sophisticated buyers and connoisseurs choose clams based on presentation. Parts of the Sound grow clams with gnarled, dirty siphons, which divers dismiss as "footballs" or "hand grenades." They fetch less money than perfect clams with gleaming ivory necks. But one can't judge a geoduck's quality until pulling it from the muck, and divers can't legally throw back unburied clams. Freed from their burrows, geoducks can't rebury themselves and are immediately devoured by crabs and flounder. To make sure divers don't take only the choicest clams and leave the lower-quality mollusks to die wasted in the mud, laws require them to keep every one they dig.

A diver's best hope is to fish an area populated by an abundance of high-quality clams. That isn't easy. Millions upon millions of clams are off-limits, either because beds are polluted or to allow already-fished geoduck tracts to recover, or simply because the area is too ecologically sensitive. Within open fishing grounds divers jockey for position, often arguing over who gets to work the best beds. The disputes lead to arguments and fistfights and conflicts among boat captains. One crew even tried to bean rivals by chipping golf balls off their boat's deck.

It didn't take long to see how things worked. Some fishermen cheated by diving illegally at night when no one watched. Others sold clams under the table. People Doug had known his whole life poached from waters he'd fished since childhood. A friend harvested twice as many clams each day as were tallied by official monitors. The rest he bagged and stashed on the bottom so they wouldn't count against his quota. Late at night he slipped out and dived with a headlamp to retrieve them. Another man, tight with local politicians, fished for and sold so many geoducks illegally he bought a home and a sports car with the proceeds. Doug told friends that he'd heard the man's girlfriend kept a diary of his crimes.

Less than a year into his business, any edge Doug thought he had was gone. So many people made money illegally collecting clams that Doug's initial business plan sounded quaint. A friend and fish broker took Doug aside and told him about acquaintances at the Washington Department of Fish and Wildlife. He said the detectives there might be interested in what Doug saw. This broker had taken previous complaints to the cops and had found the detectives responsive. They had been willing to look into poaching allegations. The broker said Tobin should share what he'd seen with a wildlife investigator—someone like Kevin Harrington or Ed Volz.

The fish broker also dropped Doug Tobin's name to the detectives. The salesman suggested the Native fisherman could help them clamp down on poaching. He said Tobin was actually quite willing.

The first thought that registered for Volz: not another informant. By the summer of 1996, Volz had gotten used to working again without Dave Ferguson. He was no longer sure he was ready to take on another snitch. Informants were like unruly teenagers. They needed handlers twenty-four hours a day. Every mole Volz had known had wanted to be his buddy. When wives or girlfriends kicked them out, they wanted to tell him their sad stories. When they couldn't pay their bills they wanted their friend the cop to loan them money. Informants had called Volz late at night when he was with his family just because they were out and a bit bored. Plus informants always thought they knew more than the police. Now that Volz had gotten his old life back he wasn't eager to give it up again so fast. But he and Detective Harrington agreed to talk it through.

Volz and Harrington worked from a satellite bureau north of Seattle, a white single-story warehouse tucked behind a Boston Market restaurant. Volz's office was a cheaply carpeted windowless square at the end of a row of pushpin cubicles. In it he stuffed bookcases with case files in three-ring binders and filled a shelf with treatises like *The Textbook of Fish Diseases*. Volz constantly redecorated, as if to make up for the soulless architecture. He tacked strings of confiscated eagle feathers next to a hanging collection of Bowie knives or taped wanted posters for serial poachers alongside a sushi chart. A few months later it'd all come down and he'd hang photographs of trophy salmon. For a while he displayed a wall-length map of the western United States, highlighting every place a road-weary cop could find a Comfort Inn.

Volz and Harrington huddled inside and hashed out the pros and cons of taking on Tobin. They couldn't shake their reluctance; the detectives were deep into their investigation of geoduck trafficking and they weren't sure they were prepared to start managing someone new, not after the handholding and scare they'd had with Ferguson. The boat explosion may not have been fatal, but it very easily could have gone another way.

Tobin also sounded like another high-maintenance guy. Volz hadn't technically ever met Doug—he could tell a few stories about seeing him at work—but other cops had, and Volz had checked him out. The fisherman's big personality was well known on the water. The detectives worried Tobin might be difficult to corral. And there was a history of bad blood between Northwest tribes and wildlife officers that might make it hard for either side to maintain trust.

On the other hand, doing nothing wasted an opportunity. Geoduck theft in Puget Sound was exploding, and Volz saw stopping it as more than just a job.

Volz had gone from high school to the army to college and then to work, hiring on as a fish cop in 1976. Early on he patrolled marine waters at night to make sure fishermen followed the rules. Over time he came to appreciate Puget Sound's delightful peculiarities. Early in his career, while patrolling in his thirty-two-foot bowpicker Volz had caught some commercial salmon fishermen working at night in a closed area. He ordered the boatmen to haul in their gear, and as one pulled up his net, an enormous fish tail broke the surface, snared in the webbing. Volz and the stunned fishermen could see only the narrow muscle linking the creature's spotted back to its tail fin, the caudal peduncle, but already it stretched the length of Volz's boat. This was no orca or blue whale, but a bafflingly large fish. Volz and the men quarreled over its identity even after a two-hundred-foot research vessel arrived to pull it free. Before the research crew could hoist the net and untangle the fish, the creature slipped its noose and plunged into the black sea.

A week later the mystery fish washed up dead near Seattle, its tail mangled and bloody. It was a whale shark, the world's largest fish, a docile, prehistoric-looking filter feeder. Whale sharks can reach sixty-five feet. Their mouths open wider than doorways and contain three thousand vestigial teeth. They typically reside in tropical waters around reefs off the Caribbean or South Pacific. No one could explain why it had turned up in Puget Sound.

For Volz, the incident summoned everything magical about the Sound, the rich mystery of the place and what it once had been. For a century the Sound's waters had provided a seafood bounty of flaky sablefish, English sole, flounder, and Pacific whiting. Fishermen snared fat chinook with nets, filled mesh cages with shrimp, and handjigged for squid on November nights. Perhaps the bonanza should have lasted forever, but by the mid-1990s, it was becoming clear it might not. Scientists had come to view the Sound as a living system influenced by every organism within it. But many of its creatures were in steep decline, from overdevelopment, pollution, and simply too much fishing.

Puget Sound was undergoing an ecological transformation, and this new order made for unusual encounters. Sea lions, once bound for extinction themselves, had made a comeback after the United States banned shooting them in the 1970s. Now they hung out near a buffet line of shipping locks, chomping the last of a dying run of steelhead. Biologists desperate to break up this snack time took extreme measures to shoo away the pinnipeds. They fired metal pellets from slingshots, shot rubber-tipped arrows from crossbows, and dropped firecrackers that exploded underwater. They even piped in killer-whale sounds to make the predators feel like prey.

Volz knew cops couldn't save the Sound, but their job description gave them incredible power to fight one problem: illegal fishing. The detectives considered their options. The shellfish inquiry they had started two years earlier had now ballooned to unusual proportions. The tips they picked up just kept getting weirder. They heard poachers cut hull sections from their boats to hide air compressors and pretended to troll for fish while divers hunted clams below. Someone marketed stolen geoducks from a Porsche, while another guy used a pay phone outside court to arrange one last shellfish buy before jail. Thieves traded clams for Vicodin and swapped them for untaxed cigarettes. It was crazy. By their calculation dozens of poachers were on the water every week, some smuggling tens of thousands of clams. The cops suspected the value of that theft probably surpassed several hundred thousand dollars a year. Later they would realize that wasn't even close.

The detectives could always use better inside information, and Tobin might fill that void. He was smart and creative and drew people to him and had spent just enough time in prison that smugglers might confide in him. He didn't drink and wasn't known to do drugs, and he would be the rare informant not trying to avoid a jail hitch. Tobin basically had just volunteered.

In their office that early July, Volz and Harrington found a compromise. They would take Tobin on but make him someone else's problem. The detectives would introduce this new informant to a federal agent—but not to just any officer. They would take him to one of the nation's best, a federal fish cop with the National Marine Fisheries Service. Special Agent Richard Severtson was an ex–Green Beret who'd served in Vietnam, a bureaucratic elbow-swinger who'd spent years working undercover. If Tobin worked for Severtson, the detectives could avoid some of the headaches. The Feds would share the information they got from Tobin so the detectives would still benefit from his insight. And an old pro would be responsible for keeping the informant in line.

Volz and Harrington could stay focused on their own geoduck investigations, which already managed to dominate their caseloads. It had been that way for so long now that most days they didn't give it much thought. But sometimes it struck them as entirely preposterous. Even with all of Puget Sound's interesting sea life—the snailfishes and tube snouts and pricklebacks and nudibranchs—this odd long-nosed clam was the thing people craved. And every poacher with a face mask thought it'd make him rich.

chapter three

CLAM KINGS

The geoduck was no stranger to thieves. Since its discovery by nineteenth-century explorers, the clam had proved irresistible to crooks. The mollusk was besieged by unscrupulous seamen and researchers even before scientists had settled on what to call it.

It began with a voyage in 1838. Under orders from President Andrew Jackson, six ships sailed from Virginia bound for the Pacific Ocean. Over the next several years, the 346 sailors of the U.S. Exploring Expedition would visit the white-tipped daggers of the Andes, Antarctica's ice deserts, and the sun-baked shores of Sydney and Honolulu. The men would gather artifacts, map coastlines, draw navigational charts, and study distant cultures. Before the ships turned toward home, the sailors would set a course for the Pacific Northwest, where scientists would collect flowering plants, coral, otter skins, shellfish, and jars of seawater and cart it all back east in the name of science.

This was North America's richest period of ecological discovery, and the ships carried men who would become some of the biggest names in science. Entomologist and museum owner Titian Ramsay Peale, son of famed portrait painter Charles Willson Peale, was the expedition's chief naturalist. Geologist James Dwight Dana would trade mollusks and argue about volcanoes and coral atolls with Charles Darwin. Also on the journey was Joseph Pitty Couthouy, a conchologist, who joined the trip after pleading with President Jackson that the hunt for valuable seashells was as important as any exploration of natural history and would pay for itself. "In conchology, particularly, it is impossible to count the rare and undescribed species which may be discovered," Couthouy wrote to Jackson. "This is a science which is daily awakening more notice in this country; and such collections might be made [which] would by their sale more than reimburse the Government for any outlay attendant upon the appointment of a person to fill that department." Couthouy argued that the allure of shells and shellfish already had been embraced in Europe, where great collections sold among the wealthy for staggering sums. If Jackson wouldn't consider sending him, Couthouy wrote, "Well General, I'll be hanged if I don't go, if I have to go before the mast!"

During the trip, Couthouy gathered thousands of shellfish from Rio de Janeiro to Samoa but eventually he clashed with the expedition's obstinate captain. Charles Wilkes was brilliant and a tyrant, and some later scholars would suggest he served as a model for Melville's Captain Ahab. He lost men to accidents, scurvy, and cannibalism; drove his sailors to the edge of madness; and would later be court-martialed for flogging and jailing crew members. Couthouy noted bitterly in his journals that Wilkes hampered his ability to collect and store new finds. Wilkes accused Couthouy of inciting a mutiny and kicked him off the expedition in Hawaii.

By the time the *Porpoise* and the war sloop *Vincennes* headed into the waters of Puget Sound in late spring 1841, the expedition was without its shell expert. The ships anchored at the mouth of the

Nisqually River, and the men fanned out to do their exploring. Some went to the water, while others headed overland toward the Cascade Mountains or the Columbia River. When the naturalists returned, they hauled hundreds of creatures back to the ships: shrimp and barnacles tagged and shoved in jars, boxes of pelts and bird skins, thousands of pressed plants, all manner of fish and many, many varieties of shellfish. From the mudflats at the mouth of the Nisqually River someone had plucked a plump and flaccid species of burrowing clam. This would become the first geoduck described by scientists—but only because it survived a rash of thefts.

No sooner had the naturalists stashed their wares below deck than the vessels' sailors began swiping the most interesting finds as keepsakes. The ships sailed back east where the precursor of the Smithsonian Institution would ultimately hold the remaining specimens. After their arrival, the collections were ransacked again. "I am ashamed to record the fact that when the boxes and packages were placed in charge of the National Institution, the seals were broken and a general scramble for curiosities took place," Titian Ramsay Peale wrote. "Many valuable specimens were lost, particularly shells and skins of birds."

Chaos among the shellfish handlers wouldn't end there. Naturalists in the field had matched notes on mollusks with numbered tin tags, which they sealed with specimens in jars of alcohol. Once in the laboratory, a priest appointed to guard the collections yanked the metal tags from the jars. "This gentleman, finding that the presence of some lead in the tinfoil tags was whitening the alcohol, carefully removed all the tags and put them in a bottle by themselves without any other means of identification," naturalist William H. Dall wrote. For a fee, the reverend let prominent scientists sneak off with a number of the rarest species. "Some of those contemporary with events have told me of the prizes secured in this immoral manner, unworthy of a true naturalist, though doubtless the temptation was great," Dall wrote. Scientists hoping to describe and classify discoveries found mountains of mismatched shells and notes. Many specimens had simply disappeared.

Eventually Dr. Augustus Addison Gould, a Harvard botanist and zoologist directed by Congress to report on the expedition's mollusks, painstakingly sorted the unusual from the common. The geoduck plucked from the mudflats at Nisqually, which Gould in 1850 had chosen to call *Panopea generosa*, caught his attention. While many Pacific shellfish retained characteristics of those along the Atlantic seaboard, the unusualness of the geoduck provided Gould with evidence that the two coasts were ecologically distinct. "Where, for instance, have we the [East Coast] analogues of *Panopea generosa*?" he wrote.

The sheer variety of marine creatures brought back by the expedition awed Gould. "The number of new species is quite remarkable," he wrote. "To the scanty list of naked mollusks previously known, [I found] additions of many new and beautiful forms." He took special note of the squid and marine shells of Puget Sound, "every one of which appear to be new to collections."

Shellfish had been central to human experience far longer than scientists like Gould knew. When humans sought a place to hunker down and wait out central Africa's glacial chill nearly 165,000 years ago, they wandered to the sea. In caves in South Africa overlooking the Indian Ocean these early people gorged on whale flesh, whelks, mussels, and other shellfish. These were mankind's first seafood lovers, and the hunger for shellfish would prove habit forming.

From the opening of the Oregon Territories, settlers and outsiders salivated over Northwest mollusks. Puget Sound's shorelines were packed with so many little necks, cockles, butter clams, stubby horse clams, and geoducks that a Washington judge in 1874 penned a song: "No longer the slave of ambition / I laugh at the world and its shams / I think of my happy condition / Surrounded by acres of clams." Timber schooners tried ferrying tender Olympia oysters no bigger than poker chips to California after gold diggers and crooks wiped them out in San Francisco Bay. (One of the thieves had been a teenage Jack London; he eventually switched sides and became a fish cop and patrolled the bay for shellfish pirates.)

For those with a taste for shellfish, geoducks were a meaty discovery. Geoducks, like oysters and other bivalves, are filter feeders. One column of a geoduck's siphon sucks down water to sort and extract nutrients. The other exhales the clean, filtered liquid. Because the creatures are so huge, all that pumping takes lots of energy, so the geoduck spends its days doing nothing, just waiting for enough phytoplankton to float by to justify the effort of eating. Between meals, the geoduck metabolizes the glycogen stored in its body, and these vast reservoirs of sugar help make the clams sweet.

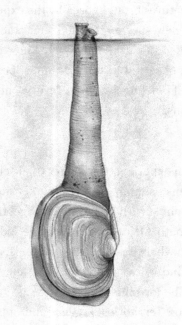

Geoduck, *Panopea generosa*

Native Americans for centuries had grubbed geoducks from Puget Sound's mudflats whenever the tide retreated far enough to collect them. They chopped the rich meat and ate it fresh or smoked and later fried up geoduck fritters.

European settlers, too, relished the giant clam. "Its flesh is, I think, the most delicious of any bivalve I have ever eaten, not excepting the best oysters," naturalist Henry Hemphill wrote in 1881. He compared the taste to that of scrambled eggs. Skillfully cooked, a geoduck would "puzzle persons who tasted it for the first time as to whether they were eating fish, flesh, or fowl," naturalist R. E. C. Stearns wrote a year later. Sliced, rolled in meal, and fried in superheated pork fat, "it would prove highly satisfactory for the daintiest epicure."

Geoducks reached epicurean heights a few years later. In 1884, the clams landed on plates before the founders of New York City's elite Ichthyophagous Club. Led by the head of New York's state fish commission and the *New York Times* editor in chief who helped expose Boss Tweed and Tammany Hall, the club brought together New York's most gastronomically inquisitive. Dining on the marine world's most unusual delicacies, these mayors, bank presidents, ship captains, and university professors hired scouts to scour oceans and streams for the most remarkable amphibian and fish species. The men bested one another by eating starfish and hellbenders (salamanders), mossbunkers (oily bait fish), and sea spiders. They laughed over geoducks and washed them down with La Tour Blanche.

But a shellfish's popularity sometimes came at a price. By the turn of the century, Olympia oysters were all but gone and geoducks had become increasingly hard to find. "There are a number of beaches that have been entirely denuded of geoducks," regional game commissioner W. W. Manier told the *Morning Olympian* newspaper in 1916. "Geoducks will be exterminated within a few years unless they are given more protection." No one knew how many geoducks existed, so state leaders set gathering limits and fined those who cheated. They made it illegal to gather the clams for sale in markets or restaurants. To make sure geoducks survived in perpetuity, geoduck collectors could only eat or give away what they dug.

It would take a folksinger with a bad mustache and a precocious vaudevillian wit to turn the Sound's shellfish into farce and make a fortune off its clams. Seattle's Ivar Haglund had spent his early years belting out songs onstage in a smooth tenor and rounding out his act with absurdist comedy. He landed gigs on radio and made friends with Woody Guthrie and opened Seattle's first aquarium in the 1930s. Ivar drew crowds with comic ditties he pretended to croon to captive sea creatures and earned publicity with stunts like wheeling a juvenile seal in a stroller to see Santa Claus. When he opened the cornerstone of a seafood restaurant empire, the name would come as no surprise: Ivar's Acres of Clams.

Ivar advertised his businesses like a P. T. Barnum of the sea. He hosted a boxing match between a heavyweight fighter and a dead octopus. When a railcar spilled a load of corn syrup, Ivar grabbed a bib and a dish of flapjacks and phoned the media to watch him ladle up the spill. When a Maine senator asked the post office to honor New England fishermen by putting sardines on a stamp, Ivar telegraphed a counterproposal: "Urge substitute bill calling for use of Puget Sound clam on stamp instead." When the government balked, Ivar sold his own four-cent stamp until postal inspectors informed him it was a crime.

His restaurants served shrimp, halibut, steak, and salmon, but Ivar made everything about the clams. Clams are slimy, enigmatic, beautiful, and disgusting, and Ivar knew that made them funnier than fish. And funny was everything to Ivar Haglund. He would produce television commercials that were both loopy and nonsensical. Cavemen chanted and circled an eight-foot clam ("Ivar's: Dancing around clams since 1938"). A time traveler in a sports car did a jig with a clam riding a unicycle. Bearded Norwegians in rain gear exchanged rhyming verses about shellfish.

Ivar liked his humor risqué and over-the-top. On a windowsill in his office, he kept a tiny plastic clam that showed a tiny plastic couple copulating inside. Ivar's personality was perfectly suited to the geoduck, which, of course, he adored. But through most of his career the clams were presumed rare, and it remained illegal for anyone, especially restaurateurs, to buy or sell them as seafood. "Because we can't sell geoduck in any form we must try to match its true glory with other clams," Ivar lamented to *Sports Illustrated* in 1964. "It's just as well it is against the law. If I ever put it on the menu, I'd start a geoduck riot. Before you know it, nobody would be eating other seafoods, everybody would be out hunting geoducks and Puget Sound would be swamped with strangers. Why, even the whole economy might collapse."

At the height of Ivar's popularity, a scuba diver's find would change everything. Through most of human history the seafloor had been a riddle, silent, wine dark, and mysterious as deep space. Few understood what lived buried in the bottom even in relatively shallow bays. The seafloor was the planet's last great frontier, and Robert Sheats was among its pioneers. He'd served as a navy diver in World War II and afterward was sent to Puget Sound, where he worked from a base that tested marine warfare systems, from underwater mines to sonar. When test-fired torpedoes lodged in the muck, Sheats and his team dived and dug the wayward missiles out. The men used hydraulic pressure hoses, which they turned toward the sand, liquefying the mud and loosening the stuck projectiles. During one excavation in 1960, Sheats and his team saw something extraordinary: a field of geoducks, the tips of their fleshy necks pushing through the seabed.

The discovery excited marine biologists and seafood aficionados. Logging and fishing still powered the Northwest, and bureaucrats saw the waving siphons as an invitation. In the years before the Endangered Species Act, the Clean Water Act, or the first Earth Day, natural wealth still existed to be exploited, and scuba-diving surveyors were sent to count and map geoducks. Tens of millions of geoducks were found in pockets throughout Puget Sound. Northwest government officials looked for ways to market this abundance, just as the region had done with its salmon and its trees. After years of research, the government agreed to lease underwater plots of geoducks to divers, like sections of forest timber sold at auction to loggers. The politicians hoped a clever entrepreneur could find something to do with these clams that would catapult them into the realm of seafood icons—Louisiana shrimp, New England cod, Maine lobsters, Washington geoducks.

Robert Sheats scored the first plot and knew just what to do. On a chilly Saturday afternoon, May 29, 1970, Sheats, his wife, Margaret, and their teenage son loaded a VW van, tied a faded black rubber raft to the top, and bumped down an old logging road past spindly pines and firs. They arrived at a splash of sand and sea on Puget Sound. They dumped the boat, threw in a gas-powered air compressor and some hose, and piled in scuba gear. They paddled a quarter mile into Thorndyke Bay.

Sheats loved this stretch of water. On clear days he could look twenty miles down Hood Canal and see his old duty station. More than that, he loved to dive. When his ship had been captured during the war, the Japanese had made Sheats and other divers retrieve tons of silver pesos that had been dumped overboard by the Allies. But Sheats hid gunnysacks in the water and secretly stuffed them with extra coins that he later slipped to fellow prisoners, who bribed greedy guards for food and radios. Later, Sheats participated in Sealab II, a navy experiment in deep-ocean survival, where he lived in a capsule at the bottom of the sea and had mail delivered by a trained porpoise. Now fifty-four and retired, Sheats just wanted to keep diving, and he wanted to do it here. Ivar had shown there was cash in clams, so Sheats hoped to start a tiny side business. His family would sell geoducks from roadside stands, and with any luck, Robert Sheats could dive forever.

They dropped anchor. With their son looking on, Sheats and Margaret strapped on tanks, double-hose regulators, and fins. In patched and marred dive suits that zipped down the front, the couple squatted on the raft's pontoons, flipped backward, and dropped thirty feet to the bottom. Sheats showed his wife how to run her hands along the floor while keeping her eyes peeled for the telltale tube of flesh. Margaret grabbed that first clam, pointed a spray nozzle in the sand, and pulled out the prize. She plucked the world's first commercially fished geoduck.

The couple hauled up fifty clams and carted them home, dumping them in the grass and across a picnic table. "We'll probably deliver some to local meat markets or seafood processors, perhaps keep a few ourselves," Sheats told the *Kitsap County Herald*. For a brief moment, the entire world's geoduck industry belonged to Sheats and his family.

Sheats's discovery might have remained a quirky footnote in Northwest history if not for a state utilities auditor with a head for numbers—and for serendipitous changes on the other side of the globe. Brian Hodgson grew up outside Olympia and had spent his childhood digging clams. He saw early on the geoduck's potential but knew that to be successful he'd need to jump-start a culinary sensation. Even before securing his first underwater clam plot, Hodgson and several partners quietly began their research.

They borrowed money, rented a warehouse, and bought a boat. They dived nonstop and yanked geoducks by the dozens. Wives and friends shucked clams at all hours. Neighbors complained about the noise and the stench. They experimented in the kitchen, scalding clams like tomatoes to float off the skins. They broiled geoducks and basted them, baked, fried, and sautéed them. They stripped off the shells and hammered at the tough breasts with mallets. They chopped clams into chowder and smoked them to create jerky. Hodgson tenderized one moist, pale, lumpy neck by slapping it between slabs of wood and backing over it in a pickup.

Even as Sheats struggled to sell his catch, Hodgson made inroads with restaurants, including Ivar's. Hodgson sold clam meat for pennies a pound to use in chowder. Late that first summer, Hodgson told the *Wall Street Journal* that his brand-new company, Washington King Clam, already "couldn't keep up with the orders." In truth, his venture looked far less promising: "We couldn't give the things away," Hodgson later confessed. "No one knew what to do with them."

Finally Hodgson asked one of his best customers, a Japanese-American seafood distributor, how she made money reselling his geoducks. She told him she took the clams to Asian groceries and sold a few overseas. Some wound up in Seattle's first full-service sushi bar. The chef there had come to Seattle from Japan's Ginza district, a region famous for classically trained sushi masters, and had spent many dawns at Tokyo's Tsukiji Fish Market, the world's largest seafood trading post. There, buyers could find everything in the sea: dried minnows and whale meat and hunks of bluefin tuna carved on band saws. Chefs who frequented Tsukiji prided themselves on using only the freshest ingredients, a habit this sushi master brought with him to Seattle. And in Seattle, his customers raved about geoducks, the crisp, springy clams that tasted fresher than everything else.

Hodgson formed a partnership with this seafood distributor in the early 1970s. He and some of his workers flew to Tokyo and marketed this clam to the Japanese. Hodgson's timing was exquisite. Personal income in Japan was rising. Shipping frozen seafood by air was getting easier and cheaper. Geoducks made an early splash. Hodgson finally started making real money, trading these durable old clams for those precious thin green strips of paper. With his success, Hodgson grew increasingly cutthroat. He bid up lease prices for geoduck beds until he drove out competitors like Sheats. For a while he would run a near monopoly.

But Hodgson lacked the connections and cultural acumen to see the next big market. The real demand for geoducks ultimately would come from a country that seemed unlikely to import anything: China. But after decades of famine, crushing poverty, and an insular economy, the world's most populous country had begun to boom, reconnecting people with their love of edible curiosities.

Like few other civilizations, the Chinese express their values through food. Lavish feasts have been part of the country's culture since at least the tenth century's Song Dynasty, but the extravagance of these imperial meals reached new heights during the eighteenth and nineteenth centuries. Rulers hosted smorgasbords packed with hundreds of eccentric dishes, from bear's paw and camel's hump to duck brains and rhino tails. That creativity lives on in modern China. In major cities, aquariums filled with snakes, scorpions, frogs, and turtles line the walls of restaurants as cavernous as shopping malls. Chefs carve meats into ornate sculptures and arrange plates of dumplings to look like flowers. Businessmen order dishes piled with a thousand rooster tongues or stacked to their chins with the meat of a hundred abalone. Popular meals are made from pigeons and sea slugs.

Claude Tchao, unlike Hodgson, understood these delicacies. In the 1970s, his Chinese father had fled the Communists for Hong Kong, where he dragged young Claude to his business meetings. Diners gorged on hearty portions of unusual dishes much as eaters did during celebrations like weddings. Few youngsters had more exposure to the specialty dishes of Chinese banquets—abalone, shark's fin, sea cucumber, fish gas bladders.

Tchao moved to the United States for college and eventually relocated to British Columbia, which by the early 1980s boomed with transplants from Hong Kong. These businessmen and families feared they would lose their assets in the transfer back to Chinese rule in 1997, and they invested billions of dollars buying land and businesses in Vancouver's suburbs. Those with strong ties to China brought with them their taste for banquet foods.

Banquet foods are chosen for sensory uniqueness, an unusual texture, taste, smell, or appearance. Sluglike sea cucumbers are soaked and boiled. Shark fin is dried and shredded in soup. Abalone and fish bladders are soaked in vinegar and deep fried. Each offers a memorable feel when sliding across the tongue. They don't look or taste quite like anything else. Tchao spent many mornings in Vancouver walking the fishing docks, hoping to find a way to make a living with his real love: seafood. After noticing Hodgson's success, the British Columbia government had opened a small fishing industry for its own geoducks in 1976. When Tchao finally stumbled across his first elephant-trunk clam, he saw in the geoduck something reminiscent of banquet food. Everything about this creature was unique—the briny smell, the fresh tang, the crunchy flesh, and the unusual proportions. The geoduck seemed extraordinarily Chinese. Tchao toured Vancouver's Asian restaurants, offering up this new ingredient. The chefs were taken by the clams and began slicing them for a type of fondue known as hot pot.

Tchao wasn't alone in seeing the potential. Geoduck samples already were being sent to a high-end Hong Kong restaurant, where the creatures bobbed in cold water in a well-lit aquarium. The restaurant served meals that could run several hundred dollars. The high prices lent the clam an air of exclusivity that impressed wealthy Hong

Kong businessmen. By 1984, Claude Tchao started shipping to Asia himself.

The geoduck arrived as Hong Kong and southern China were experiencing transformation. Between 1983 and 1993, the percentage of Hong Kong's restaurants dedicated to seafood would double. And Hong Kong sat directly across the water from one of the first areas in China to boom in the wake of economic reforms. Guangdong Province was home to new factories supplying the West: garment shops, robotics plants, helicopter manufacturers, and battery makers. The number of toy factories would balloon from a dozen in 1979 to more than a thousand just eight years later. The province's capital was a colorful center of Cantonese cuisine where diners liked to handpick their meals from bins and cages of live animals. Traditionally, only the elite could afford China's most striking foods, but soon wealth would spread, lower classes would indulge, and many trendy food fashions would arrive via Hong Kong, where geoducks in restaurant tanks already drew admiring glances.

With Seattle three hours by car from Vancouver, it was only a matter of time before Washington divers and seafood brokers found this market, too. Japanese restaurants already diced geoduck for sashimi and baked it in mushroom casseroles. Chefs in Hong Kong served geoduck with cucumber and orange, sautéed the clams with black pepper and scallions, or dropped thin slices into a wok with broccoli for a stir-fry. Groceries in South Korea sold geoduck steaks. Soon residents of several major Chinese cities would warm themselves with brothy geoduck chowder. By the late 1980s, China's explosion as a *consumer* power was just around the corner. The geoduck's future fairly shimmered. If only China's richest one percent ate geoducks, fishermen would eventually have more customers than they had clams. Hodgson's path to true riches lay before him.

But geoduck fishing remained tightly controlled, and Hodgson

already had more orders than he could fill. So Hodgson had been finding creative ways around the rules. He kept tighter tabs on regulators
than they kept on him. He ordered divers to bring up "product" by
any means necessary and paid workers to man a sixty-thousand-dollar
radio surveillance system. A friendly air traffic controller warned divers about "unannounced" airplane inspections by regulators of closed
areas of Puget Sound. One of Hodgson's employees paid area residents
five dollars an hour to tail state government employees who were
checking to see if divers surpassed quotas. When divers were ticketed
for taking too many clams, Hodgson hired them the best attorneys.
Once, Hodgson and his divers told a fussy regulator that they were
watching him and would find a way to get him off their backs. The
regulator went home and scratched his name off his mailbox.

In June 1987, a band of protesters marched outside the United
Nations in New York carrying a photograph of their bespectacled
leader: biologist Richard "Dick" Long. The marchers said Long had
told them that fishermen were illegally overharvesting geoducks in
the Northwest to supply ravenous Asian markets. Long's "Save the
Geoduck" committee called for a clam-fishing ban and promised a
summer of escalating civil disobedience until demands were met.
Days later, Dick Long revealed himself as guerrilla prankster Joey
Skaggs, a performance artist who made a career skewering the
media. He knew nothing about clam fishing or geoducks and had
caught his first glimpse of the strange mollusk only months earlier,
on April Fools' Day. Skaggs had made up everything to try a con
with a phallic prop. He duped United Press International, which sent
a story over the wire. A New York television station and a few radio
broadcasters were also taken in, as was *U.S. News & World Report,*
Germany's *Der Spiegel,* and a handful of Japanese newspapers.

Joey Skaggs, in costume as biologist Richard "Dick" Long,
holds a geoduck on the docks in Seattle.

Skaggs was giddy. "I feel so guilty," the lifelong trickster told the
New York Post's Page Six. "I swear I'll never do it again." Months later,
the performance artist would marvel at his unwitting prescience. His
joke had brushed closer to the truth than he knew.

Rumors about Hodgson overfishing had long since worked their
way to authorities. In 1979, the Washington State Patrol had secretly
investigated reports that Hodgson kept cops on the payroll. The FBI,

the IRS, and a federal grand jury investigated corruption and tax evasion allegations in the early 1980s but brought no charges. No one on
the planet understood the geoduck industry like Hodgson. He told one
friend that as an auditor he could hide financial transactions in ways
no one could unearth. "They'll never catch me," Hodgson said.

But the clam king's downfall had been set in motion before
Skaggs's skit. Six months earlier, on January 4, 1987, a light-hued boat
with a faded cabin, the *Bandito,* knocked gently against the waves in
Dumas Bay south of Seattle. Dumas Bay had been off-limits to clam
fishing because a sewage outfall drained wastewater there from a
nearby treatment plant. From a living room window, a family noticed
a fishing pole on the boat bouncing around ignored in its holder while
two men—one in diving gear, one in street clothes—scurried about
the deck. What kind of angler ignores a fish on the line? The family
called wildlife agents. A patrol boat stopped the *Bandito* and found
a few hundred pounds of stolen clams in garbage cans in the hold.
The case would have ended with a citation for fishing in closed waters
if someone had not noticed that the fishermen were Hodgson executives—one was the plant manager at Washington King Clam.

Prosecutors shuttled the case to Seattle fraud prosecutor Marilyn Brenneman. The daughter of a mechanic understood workers who
took physical risks for pay. She had taken on securities crimes and
homicides, and actually liked complexity. She empathized with divers locked out by Hodgson's monopoly. She and the lead investigator,
wildlife cop Kevin Harrington, trekked to biology labs, studied up on
shellfish ecology, interviewed divers, and scoured Hodgson's banking
and shipping records. Brenneman decided to treat Hodgson like an
organized-crime boss. Although aimed at drug runners and money
launderers, the state's criminal-profiteering laws were modeled after
federal racketeering statutes and could be used to stop other criminal
operations, seize profits, and force crooks to forfeit property. The case

that investigators put together over the next twenty months came to be known as "Clam Scam." At the time it was the largest white-collar fraud case in Northwest history.

They learned that Hodgson had ordered divers to falsify records and that he distributed a government map of Puget Sound clam beds called "the poacher's handbook." He was able to drive out competitors because he overpaid for harvest leases, then propped up his bottom line by stealing clams by the tens of thousands. He paid bonuses to thieves and threatened other fishermen. When a diver tried to union-ize, someone firebombed his boat. Harrington and Brenneman esti-mated that Hodgson had overseen the theft of $1 million in clams.

After years of being bullied, divers were quick to squeal. One had kept a record of crimes he'd committed on Hodgson's orders. When the documents were subpoenaed, Hodgson urged the man to burn them, but he refused. In 1988, prosecutors charged nine people and three businesses. The government had 180 potential witnesses and 70,000 documents. Several defendants ultimately pleaded guilty. Kevin Harrington was confronted by one defendant, and the detective never forgot the flustered man's words. "You shuffle around like you don't have much going on upstairs," Harrington recalled him saying. "But you're just like that Columbo guy on TV—always thinking." Har-rington would savor the comment for years.

Hodgson insisted on fighting his case in court. The pivotal moment in his trial came when a retired diver took the stand. In a pressed pink shirt, the diver described for jurors the scariest night of his life. The Pacific Northwest is particularly susceptible to natural disasters. Wild-fires. Floods. Earthquakes. Volcanic eruptions. Some on the jury still recalled the famous windstorm in February 1979 when winds raged so fiercely they sheared off sections of a bridge across Hood Canal. Gusts topped one hundred miles an hour. Politely and methodically the diver told the court that Hodgson had ordered him to get clams that night.

When the diver protested that the seas were too perilous, Hodgson threatened to have his job. Halfway through the evening, the terrified fisherman had radioed Hodgson that he couldn't go on. Hodgson said he'd better not leave the water without a boatload of clams.

Hodgson's lawyers stopped the trial. Hodgson pleaded guilty to trafficking, profiteering, and leading organized crime. He was sentenced to two years in prison and fined three hundred thousand dollars. Prosecutors shuttered Washington King Clam. The judge banned Hodgson for life from Washington's fishing industry.

Before Hodgson was sentenced, prankster Joey Skaggs wrote a letter to a Washington State senator: "I had perpetrated the geoduck hoax having totally fabricated it in my mind, not imagining that it could or would be true. I don't know if all the hoopla the hoax stirred up rattled the cages of the geoduck commercial harvesters . . . but I'd sure like to think so. I can't tell you how amused I am that life imitates art."

After "Clam Scam," the state reorganized geoduck fishing and strengthened oversight. Regulators bragged that these improvements transformed the geoduck industry into one of the most heavily controlled fisheries in the world. They insisted problems had been linked to one greedy businessman's monopoly. With Hodgson gone, geoduck crime would be a thing of the past.

chapter four

THE FED

Kevin Harrington shot south from Seattle toward one of the Pacific Northwest's worst eyesores: the frontage road that parallels Interstate 5 north of Tacoma. On July 11, 1996, Harrington was finally headed down to meet Doug Tobin and introduce him to federal agents. Harrington despised driving in western Washington, and this stretch always confirmed his opinion. Every time he traveled it he got stuck behind a stop-and-go centipede of brake lights and tried to ignore the barrage of billboards, hot tub showrooms, furniture barns, fast-food joints, and chain motels. It all obscured the area's remarkable natural beauty: Puget Sound lay just to the west, and Mount Rainier loomed in the east, its 14,411-foot summit floating above an ever-present raft of low-lying haze.

Harrington pulled into the parking lot of the Poodle Dog, a family-style café in the tiny Tacoma suburb of Fife, and strolled through double glass doors to the dining room. They had agreed to meet here because the nearby Tacoma Narrows Bridge led across the water to

Gig Harbor, the peninsula community where Tobin ran Blue Raven, his small seafood company. When Tobin wasn't on his boat or packing fish, he spent countless hours on this strip flirting with waitresses and wolfing fried food. The detectives had arranged several times to interview him, but their would-be informant had always been a no-show. Tobin blamed the missed meetings on mix-ups over rendezvous points, but it was enough of a slight to annoy Ed Volz. Trustworthy people didn't blow off cops. The detectives decided it might be better if the more easy-natured Harrington made the trip instead. Harrington could be the one to introduce the fisherman to the cop who would decide what to do next: Special Agent Richard Severtson of the National Marine Fisheries Service.

Tobin made an immediate impression. His bracelets, rings, and gold chains and his wild hair made it difficult for Harrington to look anywhere else. Seated together in the spacious restaurant, Harrington, Severtson, and another federal agent listened as Tobin recounted precise details about a dozen poachers. Tobin told them that fishermen were hiring small skiffs with outboards to greet them on the water so they could off-load illegal geoducks before official monitors tallied the day's take. He knew a diver who boxed illegal geoducks by the thousands in his garage. He said an antigovernment zealot who thought everyone was bugged was secretly fishing for geoduck off his speedboat and trying to get fishermen who owed him money to pay off their debts by helping him doctor paperwork. One woman had asked Tobin if he would harvest illegally and sell clams exclusively to her. In exchange she'd promised to give him a houseboat.

Tobin was a wonder in action. He spoke with authority and didn't seem to have missed much. Geoducks were worth more than ever, thanks to rapid globalization and China's growing wealth. Hong Kong and China were on their way to consuming more than 90 percent of the geoducks sold in the world. The lure of fast money was attracting new fishermen to Puget Sound, and Tobin knew the waters, the markets, and how to move top-drawer mollusks. He was serious, although funny, and comfortable with strangers. Harrington found himself enjoying the man's company. And Harrington could see that Severtson enjoyed Tobin, too.

Midway through lunch Harrington knew Severtson was sold. He would find a way to put Tobin to work. Severtson was clearly charmed by the fisherman and later admitted that he thought Tobin should have been a stand-up comic. "People would have paid good money to see Doug, and he'd leave them rolling on the floor," Severtson said. By the next afternoon Severtson's decision was formal. The National Marine Fisheries Service registered Tobin as a confidential informant: CI#9603825.

∼

By then the list of people stealing shellfish had grown so long that investigators had trouble keeping their suspects straight. The cops could name more than forty potential bad guys—only one was a woman—and had mapped the smugglers' tangled alliances with a flow chart that took four pages. Dozens of lines connected circles and boxes representing poachers, unlicensed fish dealers, restaurants, and grocers. Poachers worked the water every day. Divers fetched six dollars or more a pound for good clams, and good divers could gather several hundred pounds in a few hours. Single three-pound geoducks resold in Asia's retail markets for the equivalent of $60 to $100 U.S., sometimes more.

This seafood crime wave was gathering steam in part because similarly creative smugglers in Asia had found new ways to tap the Chinese market. Beijing had embraced high import tariffs to limit products from the West. But through the mid-1990s, according to the U.S. government, a half-billion dollars in American wildlife and agricultural products—from fresh vegetables and tree nuts, to oranges and frozen chicken meat—illegally crossed from Hong Kong to China. The United States suspected that 20 percent of the 2.5 million boxes of apples sold to Hong Kong by Washington farmers were funneled into China this way. The seafood operations were among the most lucrative.

Night after night, shipments of seafood arrived in Hong Kong from all over the world: lobsters from Australia, abalone from South Africa or the United States, and geoduck clams from the Pacific Northwest. These food items typically moved into China through its southern exotic-food capital, Guangzhou. Truck drivers in Hong Kong ferried the imports to a beach where a cargo ship moved them a few miles offshore to a rural fishing village, Kat O, also known as Crooked Island. From there, dozens of poor fishermen in speedboats motored the catches across to Chinese beaches in nearby Yantian. Efficient boaters made the trip three or four times before dawn and earned the equivalent of a month's wage in a single night. Hong Kong fishermen sometimes made enough money to hire boat pilots and lookouts and bribe Chinese customs agents. Beijing's occasional smuggling crackdowns sometimes left crates of clams dying on the dock and fishermen facing prison time—or worse—in China's unpredictable justice system. A few smugglers were killed in boating accidents or in gunfire during confrontations with Chinese coast guard officials. But the payoff was great because demand just kept rising. One restaurant seafood buyer in Shenzhen would later tell a local Chinese newspaper: "If there were no smuggled seafood the previous night, seventy percent of the restaurants would have nothing to cook the next day."

Back on the other side of the Pacific, poachers jockeyed for a chance to reach that market. Money was suddenly floating around everywhere. The wildlife cops around Puget Sound got tips about tax evasion and mail fraud. They knew that one regulator, assigned to oversee and tabulate the harvest, collected a hundred dollars each day in bribes from divers looking to avoid quotas. It was so easy to make money selling stolen clams, they heard, that a diver's biggest fear was setting the price too low. Legal divers paid a tax on harvested geoducks. Overly low prices were a sign to competitors that a diver's clams were being taken outside the law.

There were far too many cases for state detectives to track alone, so they already had been in regular contact with the Feds, which in this case meant Severtson's agency. The fisheries service oversees the health of the country's marine creatures, and the agency's law-enforcement branch is like an FBI for sea life. The agents work under the Department of Commerce, a carryover from days when fish were seen as just commodities. They investigate fishermen who take rifle shots at sea lions or boaters who harass whales. They police the import and export of everything from lobsters and rockfish to tuna and shrimp. If sharks are killed illegally anywhere in the world, their fins cleaved for soup and smuggled into U.S. restaurants, a National Marine Fisheries Service agent will investigate. If someone sells undocumented scrimshaw carved from sperm-whale teeth, fisheries agents will check that out, too.

The agents belong to a family of federal wildlife cops, all with related or overlapping jurisdictions. U.S. Fish and Wildlife Service agents police trade in endangered species and plant and animal sales on land and freshwater. Forest Service agents tackle crimes in the woods, but only within the boundaries of a national forest. National Park Service agents handle resource crimes from Yellowstone to Yosemite. Like cops with the FBI or Secret Service or Drug Enforcement Agency, most of these investigators are given the title "special agent." Each special agent shares the advantages and shortcomings of police work for the U.S government: good pay, great benefits, and maddening internal politics.

Detectives Volz and Harrington were generally suspicious of

federal cops. They usually approached the Feds for jurisdictional reasons—some crimes carry stiffer penalties in federal court. Most fish-related transgressions are simple infractions, but the crime's severity is often linked to its scope. Catching a halibut in closed waters, for example, might lead to a state citation and a fine. Catching hundreds of halibut that way and selling them nearby could lead to a felony conviction and jail time. Organizing others to poach fish and sell them abroad could land an angler in federal prison. Such smuggling can result in conspiracy or racketeering charges or charges of violating the federal Lacey Act, a powerful wildlife-conservation law prohibiting illegal buying, selling, receiving, or transporting of plants and animals across state or federal lines.

Both detectives had been burned before, pushed aside by arrogant agents who thought they knew how to do the job better. Time and again they saw Feds swoop in, mess up, insult them and alienate their sources, and then head back to a cushy office and sell the results to their supervisors as a great case or someone else's fault. Nothing got Harrington more worked up than reliving his interactions with special agents. Once, when he was helping a fellow detective investigate poaching of black cod and canary rockfish, the other detective suggested that they share information with a National Marine Fisheries Service agent. "I told him, 'If you do, I guarantee you they will take over all of the cases,'" Harrington remembered. "'There will be meetings you aren't invited to. You will have no say in the outcome. You will be out of this case.' And we were." The Feds, Volz and Harrington believed, picked battles cautiously and took only cases they thought were sure things.

Federal agents, not surprisingly, saw things differently. They complained that state detectives lacked perspective. State detectives tended to see small infractions as criminally grave insults and sometimes wanted to make cases whether or not they mattered. The state cops had spent years dealing with the same bad guys and often wore their biases on their sleeves.

Still, Volz and Harrington held Severtson in high regard, even though he could be both better and worse to work with than other federal agents. He was a brilliant cop with a long record of successes, and he was too much of a troublemaker himself to use the detectives to advance his own career. But Severtson had a big personality and an ego, and was so into the role-playing of investigative work that he tended to keep secrets just because he could. Like Tobin, Severtson could be gregarious and talk to anyone about anything. Everyone who knew him joked about his elaborate stories. Those who knew him well got to watch him fabricate them on the fly. When one colleague, Special Agent Andy Cohen, called Severtson "a pathological liar," he meant it as a compliment.

Volz and Harrington had worked with Severtson before. There was no denying his gifts. Once, the two detectives videotaped the special agent from afar while he posed as a fisherman illegally selling salmon at a restaurant's back door. The restaurant owner, standing in the doorjamb, suddenly flipped open his wallet and closed it quickly. Severtson responded almost savagely. He whipped his head about, looking terrified. The agent and the restaurant owner chatted briefly. Then Severtson started yelling. When he finally calmed down, he completed the illegal buy. Severtson was smiling when he later explained his behavior to Volz and Harrington. The suspect had tried to play a joke, pretending to be a cop flashing a badge. Severtson, not missing a beat, had simply played along. To the detectives it had been a flawless performance.

But Severtson could also alienate his colleagues. He could appear so aloof and self-righteous that many agents refused to work with him. Even those who respected him found him mercurial. Agent Cohen once turned up the volume too loud on a listening device when he and

Severtson were practicing with electronic body wires. Severtson overheard a stranger's conversation on a nearby escalator—a brief, inadvertent moment of warrantless surveillance—and blew up. "That's a federal offense!" he screamed at Cohen. "You could lose your job and go to jail!" When the National Park Service offered federal agents discounted entry into Carlsbad Caverns for a training exercise, Severtson refused on ethical grounds, even though the gesture was simply one government agency cutting costs for another. He disconnected the radio in his agency-issued sedan because the government hadn't given him a car for pleasure. Behind his back, colleagues called him "Jekyll and Hyde." Some days Severtson was their friend, other days he was something else entirely. Fellow agents only occasionally knew why.

If an agent showed insufficient interest in his job, Severtson quickly pushed the cop aside. He didn't have patience for less than full commitment. But those who showed initiative got an education. During the early days of the Internet, while younger agents exploited new technology, Severtson pushed fundamentals. He believed in shoe-leather investigating. Rather than interviewing witnesses by phone, he made his agents show up early on Sunday mornings to catch people off guard and in person. On slow days, Severtson gathered his team members in cars and made them practice tailing one another.

Severtson's disdain for administrators and protocol scuttled many of his shots at advancement. Fellow agents marveled at how often he talked his way into trouble. As an Oregon State Police officer in the early 1970s, Severtson once wisecracked to a fellow cop that his agency could shed all of its bad apples simply by blowing up the offices of the top brass. Headquarters caught wind and kicked off an internal investigation. The inquiry was short-lived—Severtson obviously had been joking—but the blemish trailed him.

But his investigative prowess was legendary, a reputation he cemented chasing fish pirates on the high seas. In the 1980s, squid fishermen from South Korea, Japan, and Taiwan had begun patrolling the open waters of the North Pacific using thirty-mile-long webs of plastic mesh, which they unspooled as deep-sea nets. Instead of chasing squid, they raked these floating death walls across the top forty feet of sea to illegally snag tens of millions of coho and sockeye salmon. The poachers flash-froze the fish at sea, stashed them in burlap bags, and offloaded them in open water to container ships from other countries. The practice hit the American salmon industry hard even though it wasn't clear if the pirated fish had been born in U.S., Canadian, or Russian rivers. With fewer salmon returning in some years, domestic fishermen already struggled to get by. And even during periods when they had plenty to catch, they fought to sell it in a market flooded with illegal fish.

Working under the assumed name Dick Frambes, Severtson secretly recorded at least fifty conversations and telephone calls between smugglers. He persuaded his boss to convince Senator Frank Murkowski of Alaska to have the government front more than $1 million so Severtson could make an undercover buy of fish. At the smuggler's demand, Severtson dropped the money in a Seattle safety-deposit box. After months of preparation, the National Marine Fisheries Service rented a refrigerated containership and met Taiwanese fishing boats two thousand miles off the mainland. After the crew started transferring the catch, a Coast Guard cutter barreled out of the mist. A C-130 transport dropped from the clouds and showered the boats with smoke bombs. The pirate boats fled, setting off a high-seas pursuit that lasted two weeks until the Coast Guard boarded the lead boat off the coast of Taiwan. Back in Seattle, Severtson stalked the head smuggler, who carried the borrowed $1.3 million out of the bank by hand in two suitcases and into the waiting arms of police. It was at the time the nation's largest covert fish operation. With Severtson's help, the United Nations eventually banned high-seas drift nets and pressured an international antipoaching consortium to patrol the North Pacific.

If anyone could direct Tobin to what the cops needed, Severtson could.

❧

The federal marine agency's law-enforcement team worked from an industrial park in northeast Seattle, one of the farthest points from salt water in the city. But the building sat a few hundred yards from the state's second-largest pool of freshwater, Lake Washington. Not that views mattered. The crew worked every day in cubicles in a windowless room. Severtson, the federal agency's ASAC (assistant special agent in charge) typed his reports in his own small office where he listened to opera or classical music through headphones as large as gun-range earmuffs. He worked with his blinds shut and his door open. On the wall he kept a sign: IN GOD WE TRUST; ALL OTHERS WE MONITOR.

It was a mongrel crew. Special Agent Cohen had transferred from the National Park Service and came from a family of naturalists; his mother was an Audubon Society board member and his father led trips to the Galapagos. He was a thrill seeker with a weakness for spy movies and gadgets. Special Agent Al Samuels was a D.C. native raised by federal cops—his grandfather worked with the U.S. Capitol Police, his stepfather with the U.S. Park Police. At twenty-six, Samuels was the youngest and the most tech-savvy. He had returned to Seattle after a beachside posting in rural Oregon, where he grew so bored he almost quit. Special Agent Dali Borden came from Washington's state wildlife agency. One of only a handful of women in law enforcement with the National Marine Fisheries Service, she felt certain old colleagues resented her advancement. She found the political maneuvering and jockeying among the Feds overwhelming but considered the agency less provincial than its state counterpart.

Within a few weeks, Severtson had a plan for a new type of investigation. Tobin would be a one-man clandestine act. He would play a version of himself and would carry a microcassette recorder in his breast pocket to tape conversations with fellow geoduck fishermen and brokers. Agents would meet him at pay phones where they might attach suction-cup pickups to the receiver so Tobin could record the calls. On rare occasions, they might attach a body wire so they could listen to conversations. The agents, led by Severtson, would work with detectives Volz and Kevin Harrington, who knew the fishermen and the buyers. Harrington, thanks to Clam Scam, had the most experience tracking geoduck harvest and shipping records. The detectives would play central roles, partnering with the Feds on some cases and chasing others on their own.

Questions lingered. Were the suspects a string of random criminals capitalizing on a hard-to-regulate market? Or were relationships organized? If so, who was at the top? The detectives knew they couldn't take anyone's word because everyone they spoke with had an agenda. Buyers called detectives and whispered that they suspected divers of poaching. Divers called to claim that buyers begged them to poach.

Now the cops had an untested weapon: Tobin. Volz and Harrington could sense Severtson's excitement. The agent loved the cloak-and-dagger part of law enforcement. He rarely skipped a chance to leave his desk and chase a lead. He thought being in the field was the reason to be a cop. Working a big case, Severtson focused like few others, his hyperalert blue eyes penetrating and unnerving. One agent called them Severtson's "hunter eyes." Severtson's plans for Tobin marked a new attempt at management. After years spent working undercover himself, Severtson would be a producer, not an actor. Agent Dali Borden would be Tobin's handler, with Severtson looking on to guide and assist. Though Severtson might steer a bit from above, the ultimate success or failure would rest on Tobin.

Severtson had great confidence in his informant's instincts. He and Tobin already seemed to share a bond, like brothers. In truth they were similar people. Both were hunters who loved to slip into the woods on weekends. Each was a chameleon and thrived on calculated risk. And Severtson saw in the informant a certain raw aptitude. Tobin's verbal intelligence and reflexive shrewdness surfaced in conversation. Severtson had seen Tobin shift direction and recalibrate his thoughts in midsentence when he thought a listener was drifting away. Tobin's hammy nature—like Severtson's own—belied an intuitive grasp of people. If Tobin proved half as effective as Severtson hoped, the team could make some exceptional cases.

Still, it was a gamble. When possible, agents would stick close so that Tobin could get what they needed without entrapping suspects. Sometimes, though, Tobin would have no choice but to work alone. Cops would not be able to monitor his every action. It made Volz uneasy. He had learned with his previous informant, Dave Ferguson, that moles could get maimed or killed, even over wildlife. Tobin could blunder by doing or saying something that would jeopardize a conviction. His involvement ought to work to the cops' advantage, but in covert wildlife work something always went wrong.

chapter five

METAMORPHOSIS: LIFE UNDERCOVER

On the surface it looks like a simple job: How tough can it be to outwit clam rustlers? Measured against cops who infiltrate drug gangs or mob snitches strung with body wire, undercover wildlife work can appear almost cushy. Chasing thieves with toucans taped to their backs, or dressing up in a gorilla suit—as one Miami agent did in 1993 to catch a buyer of black-market zoo animals—seems more like slapstick than crime fighting. But any undercover investigation demands intelligence and discipline, whether focused on cocaine rings or poachers pocketing snapping turtles. And unlike FBI agents, with their vast crews of techies, covert wildlife cops mostly work alone with little backup, even though wildlife crooks often carry guns, too.

Plus laws protecting wildlife are more bewildering than drug statutes. They vary by species and are influenced by ecology, biology, and geography. Possess just a touch of brown tar heroin and you're a criminal. But take ownership of a flock of exotic birds and you could be a wildlife thief or a farmer. Sell a dozen common garter snakes and you're no more a criminal than any pet store owner. Sell a dozen San Francisco garter snakes, an endangered reptile popular with underground collectors, and you might be part of an international smuggling syndicate. Supplying medical researchers with primates reared in captivity? No problem. But abduct the same species from the wilds of Indonesia, as one South Carolina lab did, and you may find yourself facing a federal indictment.

Wildlife crooks sly enough to skirt these laws for profit tend to make shrewd adversaries. For Tobin to pull off what the Feds had in mind, the informant would need to work within the rules but know how to push the right buttons and work his sources. In short, he would need the skills of a great undercover agent, someone like U.S. Fish and Wildlife Service Special Agent Ed Newcomer. Considered one of the best covert operators in the business, Newcomer thrives on creativity, charisma, intellectual nimbleness, and persistence. He found out early on that even the simplest cases can disintegrate. But he learned he could recover by being a great liar.

❧

Newcomer began figuring all this out one May afternoon inside a wing of the Los Angeles County Museum of Natural History. The corridors that day were crawling with bug people: entomologists, bug collectors who mounted insects in display boxes, people who kept bugs as pets, bug sellers, bug buyers, artists who painted bugs, jewelers who made necklaces from bug parts, would-be tailors who knit clothing with bug designs, and adults with children who could appreciate a good praying mantis. This was one of the world's largest annual insect shows. Folding tables formed rows of exhibitors' booths where seated men and women hawked skittering arachnids or laminated insect-wing earrings or horned beetles from Borneo shrink-wrapped in cellophane. Visitors dickered over prices for pin-mounted scorpions or leeches swimming in Tupperware. Sixty vendors attended the fair each year, and most of them knew one another well. They bought and sold their wares on an international circuit of fairs or through eBay. Every vendor knew the butterfly baron, the guy Special Agent Newcomer had come here to meet. The butterfly baron got things few could.

The fifty-two-year-old man's name was Hisayoshi Kojima— Yoshi, to his friends. Few who knew Yoshi hadn't heard that *National Geographic* often hired him to scout rain forest jungles for insects. He said he kept a Mexican boy on a two-thousand-dollar-a-month retainer netting butterflies in Honduras. He owned a house in Japan and another in Los Angeles. He bragged about bribing border offi-cials so he could sneak endangered insects out of South America. Yoshi usually got what his customers wanted, no matter how difficult or illegal it was.

If you wanted an *Ornithoptera goliath samson,* the golden birdwing butterfly found in the Arfak Mountains of Irian Jaya, Yoshi was the guy. If you needed live fist-size *Dynastes* beetles from Bolivia, Yoshi could get dozens, but they wouldn't come cheap. He once bragged that he had sold thirty beetles in Japan for ten thousand dollars each. He could find rare swallowtail butterflies from East Mojave National Pre-serve or sell *Papilio indra kaibabensis* from Grand Canyon National Park. He snickered at warnings that collecting in those places was a federal crime. Yoshi once told a prospective buyer that eight people had been arrested for gathering green and yellow swallowtails from China, and then he offered to sell one on the spot.

Yoshi always had the rarest material at the best prices. He was thought to rake in several hundred thousand dollars a year. He was the world's most notorious butterfly smuggler and was raiding the planet of some of its most endangered species.

Cops had followed Yoshi for years, but every time they got close, he managed to slip away. He regularly bragged to colleagues about outsmarting investigators. When an agent heard Yoshi would attend the May bug fair, the tip worked its way to Special Agent Ed Newcomer.

Newcomer had become a wildlife cop in his thirties, later than most. He had been an assistant attorney general for the state of Washington where he prosecuted disciplinary cases against doctors, nurses, and veterinarians. He had turned down an offer to work for the FBI in order to join the U.S. Fish and Wildlife Service. Newcomer considered the wildlife service an elite posting, one that offered more range and freedom. The FBI fielded nearly thirteen thousand agents, but Fish and Wildlife employed only about two hundred and investigated wildlife crimes all over the country.

Newcomer had dreamed of working undercover. He believed he could blend in almost anywhere. For that he credited mandatory busing. Inner-city kids had been shipped to his suburban Denver junior high. In high school, the middle-class Newcomer was bused into the city. He grew up a minority among kids of diverse backgrounds. Though wiry and unintimidating, he carried himself with confidence. He had studied martial arts since the age of twelve.

He worked under a wildlife cop who had made her name chasing alligator poachers in the 1970s. She believed in giving agents rope. Even with less than a year on the job, Newcomer believed in using it.

To catch Yoshi, Newcomer needed a fake identity. "You can't be judgmental, you can't be afraid," Newcomer once said of undercover work. "You have to open up your soul in a way. You have to buy in to the philosophy and attitude of those you're after." That would be simpler if he were also pretending he was someone else. Newcomer knew from his training that he could not map his new character too precisely. Better to merely draw in rough contours. Better to keep things vague and close to the truth. It made deception easier to manage.

Newcomer benefited from the advice of sage colleagues such as

Sam Jojola, who had been the former deputy resident agent in charge of Fish and Wildlife in Los Angeles. He kept a plastic tub beneath his desk filled with fake business cards and driver's licenses, remnants of Jojola's life undercover: a Santa Barbara real estate broker, a supervisor at a development company in Reno, an owner-operator of East Bay Antiques, a sales rep for a photocopy- and fax-machine company, an Indian trader for authentic southwestern jewelry.

Jojola once chose an identity as a phone repairman because his ex-wife worked for the telephone company. Knowing the job, the pay, and the benefits helped him infiltrate a group of marksmen illegally plugging trophy animals on the Southern Ute Indian Reservation near Mesa Verde in Colorado. Jojola spent three years on hunting excursions with no backup, no radio, and no tape recorder, tracking well-armed suspects. He sauntered off at night and scribbled license numbers or the day's highlights in a notebook stashed in his fatigues. One suspicious hunter, an off-duty policeman, even ran a check on Jojola's vehicle plates. Luckily, Jojola had licensed the car in his fake name. The hunters grilled him in the middle of the night about his identity. In town, they checked on him unexpectedly. They left after seeing his van littered with phone-company paraphernalia—key chains, coffee mugs, T-shirts.

Another operative usually adopted grumbling tough-guy personas. He wanted to be able to hang around with bad guys without having to engage in too much conversation. His fake identities gave him an excuse to rarely speak. He grew his hair long, used foul language, and scattered issues of *Playboy* and *Penthouse* around his truck. He called it "establishing my reputation as an asshole."

Newcomer knew he would need a name close to his own, one he would answer to if someone called it without warning. He also wanted a surname that started out like his own so that he could fudge a signature if he accidentally started signing his real name. This character would need to be vaguely well educated and free to come and go as he pleased. Newcomer knew enough about fishing to know the boating world included jobs few people understood. So Newcomer got a haircut and a new name. He became a middle-class suburban guy who sold boating supplies, a boring job that left time for exciting new hobbies. Like bug collecting. Newcomer called this fellow Ted Nelson.

Newcomer as Nelson had come to the Los Angeles museum that day without a plan. He went in and looked around for Yoshi. Seeing him, Newcomer decided to wing it. He strolled up to Yoshi's booth and started firing off questions, a curious guy considering bug collecting. Convincing Yoshi of his sincerity was easy because Newcomer's interest, and ignorance, was real. He hadn't had time to study up on butterflies, though he would do so plenty in the coming months. Newcomer could see that Yoshi liked the attention. He pointed out insects and quoted prices, the two men laughing and nodding. At the end of the day, Yoshi gave Ted Nelson a gift, a box of mounted butterflies to start his collection.

Within weeks the pair became mentor and protégé. Over coffee at Starbucks on Venice Boulevard, Yoshi talked about the time Mexican customs agents caught him with two hundred live beetles and the time a South American official saw a beetle's horn poking out of his carry-on. Yoshi told Nelson he sold antiques through Sotheby's and had once collected fighting fish. He said that he had run a travel business and had passports in different names—one Japanese, one American.

Yoshi explained the extremes to which bug collectors went for their obsession. True insect lovers didn't just catch butterflies but also gathered larvae, pupae, chrysalises, and cocoons. They grew special plants in their homes and reared their own butterfly specimens. The finest collectors did not want butterflies that had ever flown; flying

could scratch the delicate, papery wings. True collectors used cater-
pillars to produce new specimens before their eyes. The moment butterflies emerged and their wings filled with blood, collectors slipped
the fresh specimens in glassine envelopes and refrigerated them. The
dying insects would metabolize their fat to keep warm, and once the
butterflies were dead, the collectors could stick them with pins and
mount them under glass in display cases.

Newcomer filed away each tidbit, not knowing what might prove
useful later. He told Yoshi that he reminded him of Indiana Jones. The
balding bug collector appreciated the comparison. Yoshi confided that
he wanted to take his business in a new direction. He wanted to sell
butterflies on eBay but feared his written English was too poor: "I can
do so many things, but I could not use eBay," he said, chuckling. But if
Yoshi supplied insects, Ted Nelson could write descriptions. The two
new acquaintances could become business partners.

Newcomer suspected his new friend wanted someone he could
feed to the cops if he got caught. Ted Nelson, of course, agreed to help.
From the beginning Yoshi made clear his business wasn't always legal.
Yoshi warned Nelson to avoid customers demanding proper documentation. If federal agents contacted him, Nelson was to say his boss kept
all paperwork in Japan.

Butterfly trading is controlled by the Convention on International
Trade in Endangered Species, an agreement among more than 170
countries. CITES was designed to keep the plant and animal trade
from overexploiting rare creatures, and it protects thirty thousand
plants and animals using three categories, reflecting the varying levels
of scarcity and risk posed by commerce. CITES III species generally
thrive despite trade, but purchases are regulated and require export
permits just in case. CITES II species do not yet face extinction but
could, so nations limit exports to stabilize populations. These species
are carefully controlled, like prescription drugs. Buying or selling

CITES II species isn't illegal unless done without permits. CITES I species, except in rare circumstances, are vanishing so rapidly that commercial trade is outlawed. Getting these species was Yoshi's specialty.

Newcomer went on bug-trading Web sites and found that buyers posted thoughts about one another. It seemed like a good way to ingratiate himself to Yoshi. Ted Nelson would post a message vouching for the butterfly man and start taking requests for online orders. Newcomer thought Yoshi would appreciate the initiative.

Yoshi instead excoriated his partner. Ted Nelson didn't know what he was doing. Ted Nelson had moved too quickly. Ted Nelson did not properly screen his customers. Ted Nelson was rash. If the feds raided Ted Nelson's home, they'd find Yoshi's name on his computer. Ted Nelson was going to get them both caught!

Newcomer had overplayed his hand. After a while Yoshi returned to Japan. The men stayed in touch, but the relationship grew distant. Yoshi was gone, and the agent had not witnessed a single illegal transaction. Newcomer decided to have Ted Nelson sell his own butterflies on eBay, advertised with digital pictures Yoshi had given him. At the opening of each auction, Newcomer would assign other federal agents to post the highest bid. Newcomer held dozens of online auctions, hoping to illicit a response from Yoshi.

He didn't expect the reaction he got.

Yoshi once again was livid. Ted Nelson was now competition, and Yoshi started campaigning against him. Whenever Ted Nelson posted a butterfly for sale, Yoshi posted the same species on another Web site. Sometimes Yoshi added a jab. "Shame on you Ted Nelson," he wrote once. "You're using my photos without permission. You don't have CITES for this." Yoshi advertised his goods as cheaper "than eBay auction and Ted Nelson."

Newcomer's aggressive tactics weren't working. More than a year after their first meeting, Newcomer had actually turned Yoshi against him. One day Newcomer took a call from the California Department of Fish and Game in San Diego. A Japanese man ranting about butterfly smugglers—one smuggler in particular—had left a recorded message on the agency's tip line. Yoshi, calling anonymously, had turned Ted Nelson in.

A long line of men and women have gone to similar lengths trying to catch people committing wildlife crimes.

Before the creation of the CIA, the DEA, or even the FBI, some of the country's sneakiest covert work involved wildlife smugglers. Agent Phillip Farnham of the Bureau of Biological Survey went undercover during the Great Depression to round up fur importers smuggling silver foxes from Canada. In Chicago, federal agent John Perry transformed himself into a drunken hobo named Dopey who wore stained trousers, scuffed boots, and natty sweaters. He tracked bird thieves who illegally supplied street vendors on the city's waterfront. Smugglers paid Dopey to ferry packages among vendors, recognizing their error when Perry strolled into court in his dress uniform. Another agent in the 1930s worked undercover as a drummer in a swing band and caught duck bootleggers who hid contraband in nightclub iceboxes. In the 1960s, a covert agent code-named Peanut Man dispensed nuts in bars and restaurants in Texas and Louisiana as a ruse to spy on bird traffickers. He arrested so many on one outing that the government took them to jail in a school bus.

In the years before World War I, even the crime-busting Pinkerton Detective Agency's private eyes went after animal bandits. According to Louis S. Warren's *The Hunter's Game,* two years before the creation of the FBI a Pinkerton known only as Operative 89S lived incognito for more than a year with a gang of bird poachers while he hunted the thugs who had killed a Pennsylvania game warden.

Newcomer knew his case was in trouble, but he insisted on salvaging something. He asked another agent to pose as a collector and purchase butterflies from Yoshi online. In a moment of weakness the careful dealer let down his guard and sold this stranger three Bhutan Glory butterflies, without permits. A package of butterflies arrived from Yoshi. They were smuggled, misdeclared, and lacking proper paperwork. The agents had finally caught Yoshi committing crimes.

But the crimes were a joke, and Newcomer knew it. Rather than a conspiracy to smuggle thousands of imperiled animals, the agent, after more than a year's work, had caught the butterfly kingpin trafficking $137 worth of insects. It was like nailing Pablo Escobar for snorting a line of coke. No U.S. attorney would touch a case so small. Worse, Yoshi had stopped returning Ted Nelson's e-mails or posting on Web sites. Newcomer ran several more eBay auctions, but the butterfly king didn't surface again.

Newcomer had torpedoed his own case. The agent figured he'd never see Yoshi again.

The Yoshi files sat untouched on Newcomer's desk for two years while another case ballooned around them. A band of Los Angeles birders were breeding roller pigeons, birds with a bizarre genetic tick that makes them spin backward and plummet to earth in midflight before righting themselves. The birds' acrobatics were a spectator sport, with breeders judging flocks of twenty, called kits, based on the quality and synchronization of their rolls. Unfortunately for the pigeons, the tumbling drew predators that thought they were weak or injured. Breeders responded by killing the predators. They trapped and stomped on Cooper's hawks. They suffocated red-tailed hawks with sprays of bleach and ammonia and gunned down peregrine falcons with air rifles or shotguns equipped with silencers. One birder claimed that he killed forty each year. Another filled a five-gallon bucket with talons. The breeders as a whole were killing two thousand birds of prey a year, species protected by the Migratory Bird Treaty Act. And peregrine falcons, already hammered by pesticides, had only recently staged a comeback.

Newcomer resurrected Ted Nelson, this time as a struggling, not-so-smart blue-collar guy. The new Ted Nelson was down-and-out. He wore ratty jeans and kept his hair thick and grimy. A fat biker mustache curled around his chin. He worked for a boat builder and drove a crappy van. He had so little money, he told people, that his boss let him live in the company warehouse.

Nelson infiltrated competitions in which birders traveled house to house, from the San Gabriel Valley to downtown Compton. Sometimes he wore a wire; sometimes he shot video from a tiny hidden camera. He met a man who built and sold hawk traps in a parking lot behind a store that sold pigeon paraphernalia. Leaning against a backyard porch was a .20-caliber pump-action air rifle. "Five pumps of this," the owner told Newcomer, "hawk's gone."

A few weeks into his new operation, Newcomer got another tip. Yoshi was back and headed to the bug fair after skipping the last two events. Newcomer decided to engineer a run-in. Ted Nelson could apologize and clear the air. Newcomer had to play his role perfectly. He stood at the far end of the museum and filmed Yoshi from a distance with a zoom lens. He watched Yoshi move from table to table in his flower-print shirt, dark slacks, and tennis shoes, chatting up fellow bug lovers. When Yoshi walked toward a narrow hallway, Newcomer saw his chance and headed the same way.

"What are you doing here?" Newcomer said.

"No, what are *you* doing here?" Yoshi replied.

The men shook hands and said hello. Nelson said he was glad to see Yoshi. He had, in fact, been meaning to thank him. Ted Nelson explained that someone had called in an anonymous tip about him selling butterflies illegally. Not knowing what to do, he had fallen back on advice from Yoshi: When the cops showed up, Nelson didn't let them search his place. He had also kept his best stuff elsewhere. Yoshi's wisdom had kept him out of prison, and Nelson said he was eternally grateful.

Newcomer saw that Yoshi believed him. The two men agreed to meet for lunch that afternoon. Two hours later over cold soup and cabbage at a Korean barbecue, Newcomer asked about Yoshi's health. Yoshi commented on Ted Nelson's new biker mustache. It almost seemed like the butterfly man was flirting. Both confessed they were still deep into the bug trade. Newcomer had his hidden audio recorder running.

"You ever get these *chimaeras*?" Newcomer asked. With wingspans that rival those of small robins, the South Pacific *Ornithoptera chimaera* flutter above rain forest canopies feeding on the nectar of high-sprouting flowers.

"*Chimaeras?* I think I have about ten pair," Yoshi said.

"What do you want for those?" Newcomer asked.

Yoshi was silent. "*Chimaeras* come from Papua New Guinea," he said. "You can get them from Indonesia easy. But in Papua New Guinea . . . difficult. I give them to you for . . . seventy dollars or eighty," Yoshi said finally, and then laughed.

Yoshi ordered greasy beef and pork dishes for them both. Newcomer, a strict vegetarian, forced it down.

"I might ask you to sell me those *chimaeras*," Newcomer said after a time. "This would be for me. And then I'll resell them to someone else."

Yoshi said he could send them by mail.

"What about customs?" Newcomer asked. "Will they check it?"

"Express Mail, no check."

Newcomer asked if the *chimaeras* would have permits.

"No permits." Then Yoshi corrected himself. The butterflies *would* have permits; they just wouldn't be real.

Newcomer agreed to buy all ten pairs.

Yoshi loosened more. He volunteered to get *Ornithoptera alexandrae,* the world's largest butterfly. It was the Holy Grail for collectors. With the wingspan of a football and exquisite, iridescent yellow-blue-green ring patterns, the Queen Alexandra's birdwing is one of nature's most endangered and spectacular creatures, found in the rain forests of Papua New Guinea. Alexandras are banned from all trade, but Yoshi said he regularly shipped them from Papua New Guinea to Europe and then on to Japan and the United States, covering his tracks by mislabeling the packages. "We write down that it's a moth," Yoshi said. Customs officials "don't know better."

Queen Alexandra's birdwing, *Ornithoptera alexandrae*

Newcomer's chest pounded. Maybe he hadn't blown this after all. "I wonder if any of my customers would buy an Alexandra?" he said, as if to himself.

"You be careful," Yoshi warned. "You may lead them back to me."

Newcomer backed off, but Yoshi told him to bring the next day a list of species that he wanted.

After lunch, Yoshi asked if Nelson could give him a lift to a Japanese sauna. In the car, Yoshi talked casually about other butterflies from faraway lands, which he also said he could find ways to bring in. He also spoke openly of the men who frequented this sauna, men who sometimes propositioned one another in front of him. "That right?" Newcomer said, again trying to sound disinterested. He let the moment pass but made a mental note. Men visited this sauna looking for sex. The revelation implied an escalating bond of trust. After three years of trying to put Yoshi away, Newcomer was suddenly making quick progress.

"Have fun in there," Newcomer shouted, teasing, as Yoshi stepped from the car. He had heard and understood Yoshi, but Newcomer wanted his reaction to seem ambiguous.

Yoshi laughed and headed inside.

Newcomer was thrilled to be back in Yoshi's good graces, but the agent now balanced two operations, alternating between different personalities in different underworlds. Yoshi was frumpy and middle-aged and had just confided a romantic interest in men. He was manipulative, frequently suspicious, and smart. Ted Nelson had to remain gentle and well spoken. If Yoshi suspected anything, he'd vanish in a flash.

The pigeon breeders, on the other hand, were a menacing pride of lions. Some were ex-cons who carried weapons or made their money dealing coke. One was wanted in connection with a rape, another bragged about fights he had won in an L.A. jail. Another sported gang tattoos across his back. If these guys figured Nelson for a cop, they would discard him in a ditch.

The day after the Korean barbecue they met again. Yoshi pulled a small, clear plastic box from a fanny pack with five cocoons inside. *Papilio indra kaibabensis,* from the Grand Canyon. Yoshi wanted to know if Ted Nelson could get more. The *indras* could earn them a fortune in Japan, he said. Newcomer couldn't believe his luck. Yoshi apparently trusted him fully.

Newcomer returned that weekend to San Bernardino and the pigeons. The following Wednesday, in an online video conference on Skype, Yoshi promised to send the Queen Alexandra from Japan. That weekend, Newcomer hit a new bird show. Sometimes he would spend all day talking about killing hawks, then head home and grill Yoshi about bugs while Yoshi made lewd comments about sex.

The identity shuffling began to take a toll. Newcomer kept a checklist on his dresser to remind himself of what to wear or carry, depending on which version of himself would surface that day. If his undercover phone rang while he was relaxing with friends and family, he'd change his voice pitch and tempo and talk illegal activities as Ted Nelson. At times he felt dirty and a little guilty, as if he had actually done something wrong. Then he would hang up, and friends and family would expect him to snap back to Ed Newcomer. Once, driving with his mother on vacation in Colorado, Newcomer pulled over to take an undercover call. Newcomer could see that his coarse gang talk worried his mother. The strain spread to Newcomer's wife, who expressed frustration that her husband always made time for the suspects.

Ted Nelson, on the other hand, was making great progress. Before leaving for Japan, Yoshi had looked over Nelson's list and agreed to supply every species on it. During a video chat, Yoshi held endangered

Corsican swallowtails up to the camera from Kyoto. He would sell six for seven hundred dollars a pair, even though they were banned from trade. Yoshi confessed that his personal butterfly inventory now topped a half-million dollars. He also promised he would send Newcomer endangered peacock swallowtails. In exchange, Yoshi wanted Arizona maps. He planned to mark areas in the Grand Canyon where Ted Nelson should start collecting. Then he offered Newcomer the rare hybrid Southeast Asian butterflies *Ornithoptera allotei*. The price: thirty thousand dollars.

Newcomer's packages began rolling in. Six weeks after their reunion at the bug fair, Nelson had bought twenty-six thousand dollars' worth of illegal butterflies. Yoshi had offered three hundred thousand dollars more in merchandise, and now Newcomer had digital video of Yoshi committing felonies. He faced one last hurdle: his suspect was in Japan, with no immediate plans to return.

The smuggler ultimately provided the solution. He grew bolder the more the men talked, which was now almost daily. He openly acknowledged an attraction to Ted Nelson. Yoshi made lurid comments when dickering over prices. He asked Nelson to remove his shirt when showing cocoons to the camera, and though Newcomer tried to steer them back to butterflies, an idea began to percolate.

The roller pigeon case would go on for months. Newcomer eventually set up hidden cameras outside a breeder's backyard, and fellow agents pawed through the man's garbage. In a white plastic bag they found a dead hawk dripping blood. Newcomer sent the carcass to Ashland, Oregon, where the U.S. Fish and Wildlife Service operates the world's most sophisticated wildlife forensics laboratory. Scientists identified the species as *Accipiter cooperii,* a Cooper's hawk. They determined it had died from "massive blunt force trauma to the head and spine."

In a recorded telephone call the next day, the breeder told Nelson he never shot hawks, preferring instead to pummel them with wood. "You'll see," he told Nelson. "You get a lot of frustration out." When Nelson asked how he pulled it off, the pigeon man happily explained in detail. After catching the hawk in a trap "you can just open it slightly, about three inches, just get a stick in there . . . Once you ping them one good time they're going to be somewhat dazed and then you can just go for it." Hidden-camera photographs later showed the breeder marching toward his backyard hawk trap carrying a stick.

Federal agents eventually arrested seven roller pigeon suspects. They mostly faced small fines. Killing birds of prey is a felony under the Migratory Bird Treaty Act only if it accompanies intent to sell. The case would lead some in Congress to pursue changes to make intentional and malicious bird-slaughtering a felony.

Newcomer brought Yoshi down, too. A month into their regular video chats, Newcomer saw his opening to coax Yoshi back from Japan. During one conversation, Yoshi groused that Ted Nelson owed him too much money. Newcomer made his play, grinning into the camera. If they could meet in person, Nelson would "make it up to him." He let the double entendre linger.

"Really?" Yoshi said, drawing out the word.

"You'll just have to wait until you get back to L.A.," Newcomer said.

"You're a tease," Yoshi muttered.

Newcomer laughed. "How else am I going to get you to come here?"

Two months after reuniting at the bug fair, Yoshi landed at Los Angeles International Airport. Federal agents greeted him. Six months later, he pleaded guilty to seventeen criminal charges and was sentenced to almost two years in federal prison. Newcomer stayed in hiding during Yoshi's arrest but visited the smuggler in jail the next day. Yoshi saw the handcuffs, the badge, and the empty gun holster and asked Newcomer: You're an agent?

In the end most of what Yoshi said had been lies. He had never held two passports and he never worked for *National Geographic*. Once, Yoshi had called to say he was in the United States, offering details of his stay in St. George, Utah. Two days later, an informant confirmed the story for Newcomer: He'd heard that Yoshi was collecting just hours outside St. George in the Grand Canyon. Newcomer discovered that neither claim was true. Yoshi had been in Kyoto the whole time.

The butterfly king had created a thoroughly coherent world of lies. He got caught in part because he let down his guard—and because his adversary ran a better con.

~

Tobin shared Newcomer's talents. He knew how to cater to an audience. Seniors drifting by his vessel during sightseeing tours would see Tobin the Fisherman, the coarse bad boy, turn his back and dangle a geoduck from his waist so the gangly siphon drooped to his knees. While elderly tourists gasped, Tobin would squeeze the clam's neck and make it ejaculate water onto the deck. Tobin the Carver spoke like a mystic, cloaking discussions of Salish masks and totems with such reverence that art aficionados begged him to whittle pieces. Then there was Tobin the Storyteller. One earnest acquaintance asked Tobin to visit his home and offer it a Native blessing. Tobin the Informant cackled when he recounted the scene to Special Agent Rich Severtson and Detective Ed Volz. He bragged he'd made up spiritual gibberish on the spot.

Severtson wanted Tobin to get close to Gene Canfield, a hardworking fisherman with a mop of straw hair, a houseful of sons, and a powerful libertarian streak. Canfield knew the industry's secrets. He'd grown up on the Washington coast, had been a diver with the navy, and started gathering geoducks in the mid-1970s. Canfield loved the adrenaline rush of diving. He'd fished for sea cucumbers and gathered abalone in Alaska but always found his way back to geoducks. "When you're down there diving you see all this money just lying on the ground," Canfield once explained. "Not only do you see it, but you know that you have the ability to get it out. Not only can you get it out, but you can get a lot of it out in a short period of time." He spoke of breathing underwater as "cheating nature." Federal agents suspected he cheated in other ways, too.

Canfield had told Tobin about a California clam market that would take improperly collected geoducks. The packages just had to be boxed nicely and include counterfeit documents. Canfield suggested a friend could do the deal, one who claimed to control most of the world's geoduck market. This was exactly what Agent Severtson wanted—a connection leading straight to the top illegal geoduck dealer.

Late on a hot August night in 1996, Tobin telephoned Canfield as an agent listened in.

"Oh man, I just got in," Tobin said. He had been out netting fish and said he planned to smoke them.

"For what's-his-name?" Canfield asked.

"No," Tobin said, not missing a beat, "for Doug."

He steered the conversation to complaints about a fellow fisherman who illegally sold inferior clams masked as high quality.

"The product is no good," Tobin said. "It's shit—I've seen it.

"Something else I'd like you to put some thought on," Tobin continued, hesitating, "is, ahh, uhm, I trust you there, so I wouldn't tell anybody this, but this . . . it's completely wide open out there." As a one-man operation, Tobin said, he could get several dozen tons of wild geoducks. If Tobin poached in large volumes, could Canfield sell it all?

Canfield told him to just box the clams and get them to the airport.

He could keep their hands clean by finding a market for the clams with another company.

"Right," Tobin said. Canfield had said exactly what the agents wanted to hear. Tobin tried to get Canfield to be more explicit. "You know, that company you had down south, whatever it was . . ."

"Yeah, yeah," Canfield said obliquely. "He'll do it."

Tobin scratched and prodded a bit longer, looking for dirt. When he sensed he had pushed Canfield too hard, he slid back into banter. He talked briefly about carving and asked Canfield if he had delivered one of Tobin's prized masks to an acquaintance. He complained about a rival fisherman who stole geoducks at night. Canfield shrugged it off, suggesting Tobin should do the same. They talked about the organized criminal gangs that everyone had heard controlled geoduck smuggling in the People's Republic.

"In China?" Canfield asked. "Yeah, I believe that. But I don't think it's the Mafia."

It was close. The more China's economy exploded, the more corruption flooded in with the dollars. Hong Kong thugs ran protection rackets like the Italian Mafia and demanded bribes from seafood importers even in China, threatening violence to those who didn't pay. Gangs sometimes even ran the police and leveraged influence with businesses and the political elite.

Tobin told Canfield that sooner or later it would make sense if he could make a trip south and meet this secretive broker himself. Canfield said he'd work on making the arrangements but first he and Tobin should get together and talk. They agreed to meet later that month.

Less than a week later, Tobin slipped out to Canfield's house on a dead-end lane in the secluded central Puget Sound village of Olalla, across from the western shores of Vashon Island. Inside the two-story blue-and-white farmhouse, Tobin got to work. Canfield tutored him in illegal geoduck harvesting, suggesting that he properly document only half of what he harvested. Canfield's contact in Las Vegas would direct Tobin where to send five hundred pounds each day. He said that Tobin should sell the rest legally. Canfield's broker would wire six dollars for each pound shipped, though Canfield would subtract a dollar per pound in finder's fees. Still, it could mean as much as three thousand dollars a day for six hours of work. Tobin taped it all on a hidden black Panasonic microcasette recorder he called his "little black buddy."

These conversations were a pipeline to the underground markets. It's what the agents had been missing. With Canfield, Tobin found his rhythm, intuitively sensing how to steer conversations. Severtson was pleased, but to make solid cases, the cops would need to document every step of each illegal act several times. Poachers often got off lightly, claiming they didn't know the rules.

The next day, Severtson opened an account in Tobin's name at Seafirst Bank. Later, over dinner at Pearls by the Sea, Tobin told Canfield that he had collected 1,347 pounds of clams that day but had declared only a bit more than half on his quota. The men agreed they would ship the excess that night to buyers recommended by Canfield's Vegas friend. They talked more about divers and boats, the timber industry, horse clams, and an upcoming Squaxin Island tribal council meeting. Canfield warned that Tobin should keep cash moving in and out of his bank account. The bank would be the first place suspicious state or federal agents would check.

That night, Severtson, Borden, and Bill Jarmon, a Fish and Wildlife detective, met in a warehouse and packaged 410 pounds of geoducks in eight shipping boxes with Tobin's contact information. They put no marks on the outside, as Canfield had instructed Tobin. National Marine Fisheries Service special agent Al Samuels, the most computer savvy, used a graphics program to make a label for Tobin's company, Blue Raven. Severtson took the packages to Sea-Tac International Airport and left them in a cargo truck, which would ferry them later to a waiting plane.

The next day the Las Vegas broker called Tobin directly. He complained that the labels on his packages were incorrect and made it clear that he was not happy. Federal agents might notice. Future shipments needed the date and location of the harvest stamped on the outside. Nevertheless a wire transfer arrived later that day for twenty-four hundred dollars.

Canfield had led Tobin and the cops to the geoduck kingpin: Nichols P. DeCourville.

chapter six

KINGPIN

The detectives knew the name. Six months earlier while investigating a small seafood shop, a state health inspector had found illegal pinto abalone for sale in the merchant's tank. Detectives came and riffled through this South Korean seafood dealer's black address book. He admitted he'd been buying tens of thousands of dollars' worth of bootleg clams from a diver named Henry Narte, who was already under investigation by Ed Volz and Richard Severtson. This dealer told the detectives that Narte and his crew paid lookouts $150 apiece to make sure no one saw them fishing from secret, illegal spots so rich in clams the spots had nicknames: "Shell City" and "Swiss Bank." The men sometimes worked the water for nine straight hours, bringing up twenty thousand dollars' worth of geoducks in a day. One crew member bragged that they had made a quarter-million dollars in eight months. They bought cars, stereos, jewelry, and televisions. The lead diver owned a Corvette and a Mercedes and negotiated on the docks from the seat of his black Porsche. Most interesting to detectives: The

South Korean dealer resold most of these clams to a seafood broker in
Las Vegas named Nichols P. DeCourville.

DeCourville had contacted the cops several times in years past
to complain about rival dealers. Federal agents in Los Angeles had
even seized a shipment of clams from Washington with an invoice
from NDC, DeCourville's company, because its paperwork was not in
order. The documents on the shipment belonged to a company that
had folded years earlier. At the time it was not clear if it was a minor
case of fraud or a simple mistake. Then that spring while researching
records at Sea-Tac International Airport, Special Agent Severtson saw
a box that read: FRESH SFD. Federal law requires that seafood be identified by species, either on the outside of the box or in accompanying
documents, so Severtson called the number on the container. He was
connected to Nichols P. DeCourville, who owned up to the blunder
and promised future compliance. The guy had set Severtson's antenna
quivering. Now it appeared DeCourville's clerical errors added up to
something more.

The investigators learned about DeCourville slowly. He'd led an
unorthodox life. A bout of polio had put him in leg braces as a child and
left him with a slight limp. When he was young, his mother had kicked
his father out, and she forced Nick to deliver papers and set pins in a
bowling alley to help pay the rent. Nick left home as a teen, hitchhiked
to West Virginia, and paid his way running errands for prostitutes in
the Ohio River brothels. By sixteen, he was washing dishes in a bar
and playing poker on the side. When a drunk pulled a gun on him
during a card game, DeCourville finally split. He doctored his birth
certificate and joined the navy, which sent him to the commissary at
Pensacola. DeCourville ran the kitchen and found his true calling:
food. He got out, married, had two sons, and became an accountant at
an auto-supply outfit. He divorced, remarried, and got divorced again,
then married and divorced a third time.

DeCourville eventually worked his way into running an elegant seafood restaurant and dance club on L.A.'s Sunset Strip. He also co-owned a Beverly Hills Porsche dealership, but the restaurant, Nick's Fish Market, was his real love. It drew celebrity guests like Chuck Norris, Casey Kasem, and Stevie Wonder. DeCourville's obsequiousness kept them coming back. Every day at 5 A.M. he hit the fish market in downtown Los Angeles, selecting only the freshest creatures. He ordered waiters to arrive early so they could call diners about their evening's special requests, and he insisted his celebrity chef make rounds among the patrons. A scribe trailed the cook, jotting down suggestions. When Hugh Hefner hosted a party to close the original Hollywood Playboy Club in 1986, DeCourville promised "to keep the memory alive" by hosting former bunnies at weekly parties at the restaurant.

One DeCourville patron was an attraction in his own right: Joseph Isgro, a record promoter the FBI suspected of having mob connections. Years later Isgro would face more than fifty federal charges of money laundering and racketeering, all of which would eventually be dismissed by a federal judge. A decade after that he would plead guilty in a new extortion case. But during the mid-1980s he drew so many undercover cops to Nick's Fish Market that DeCourville's friends called the restaurant the safest spot in L.A.

By then DeCourville had remarried again, this time to a beautiful concert pianist who spoke four languages and claimed to possess the world's fastest fingers. She would later travel the world as "The Piano Princess." But at twenty she drove a white convertible Rolls-Royce and played alongside Liberace. Though she found DeCourville charming and handsome, she was a quarter of a century his junior and a bit naïve. She learned after they married that DeCourville faced serious problems with the IRS and, he told her, the Mafia. In public, DeCourville pointed out mob figures to her. There was a phone in their closet that he told her could not be tapped. He claimed he was an informant for the FBI and told her he used the phone to pass along tips about gangsters.

The couple divorced at the end of the 1980s and DeCourville took a job as a seafood broker, which required him to travel to Asia and buy load after load of shrimp. Then he moved to Vegas to start NDC, through which he traded shellfish, particularly geoducks, over the phone. By 1996, he brokered much of the nation's geoduck supply and made deals for customers all over the world.

Gene Canfield and Doug Tobin kept talking regularly. They spent hours on the phone and met in person every few days, either in restaurants or on the water. They talked about how best to poach clams. Canfield told Tobin to harvest during the day and then leave the clams tied in net bags on the seafloor. With a GPS reading, Tobin could return at night to retrieve the clams with grappling hooks. He showed Tobin how to use radar to determine if anyone was on the water nearby, and he told him about running an air hose straight through the hull to conceal divers. If anyone approached at night, Tobin could simply lift the anchor and drift away from the evidence.

But after that first phone call when DeCourville had yelled at Tobin, the informant talked regularly to the Vegas kingpin, too. By Labor Day, Tobin was helping the cops funnel seafood through Canfield to DeCourville's customers. The Vegas broker quickly warmed to his new supplier. The men already were making plans for Tobin to come to Nevada, where Nick hoped they would celebrate their new business partnership.

"You let me know when you're going to be in Vegas and I'll take care of you," DeCourville said. "You got a girl or a wife or something? Bring her with you."

"I got a girlfriend," Tobin said, "but if I don't bring one, I'm sure you've got one down there for me, don't ya?"

The men frequently didn't bother to include Canfield in their plans. Within days of getting DeCourville on the phone, the informant had begun worming his way in between the Vegas broker and his loyal libertarian diver. In early September, DeCourville called Tobin asking for five hundred pounds of clams in a hurry as "a personal favor." Tobin tried to draw the buyer out. He asked DeCourville if he was sure that Canfield could be trusted.

Canfield would "never rat anybody out," DeCourville said, though he was convinced that others had spoken to the FBI. "You're not going to believe this when I tell ya," DeCourville said. He named another buyer both men knew.

"Is that right?" Tobin asked.

"That's right."

"I'll be goddamned."

"Yeah, he brought them in last year to get rid of his competition . . . called the FBI," DeCourville said.

"That sleazy piece of shit," Tobin said.

"Yeah he is."

"I hate a fucking rat," Tobin said.

DeCourville laughed. "I do, too."

During these conversations, DeCourville explained bits and pieces of his operation. Tobin learned that DeCourville connected fishermen with buyers, but that the goods never went through Las Vegas. DeCourville made a point of never actually touching the clams. That way he could distance himself from any crimes. If one of his suppliers actually got caught poaching, DeCourville could shrug and blame the reckless diver. Once, Tobin asked DeCourville outright if Canfield had explained all the illegal ways they got his clams.

"You know what?" DeCourville said. "Yeah, he did, but I don't give a shit. I don't even want to talk about it."

DeCourville confided that he controlled 70 percent of the domestic geoduck market. He said he bought from a variety of sources and moved two thousand pounds or more of clams per day. Depending on demand they might sell for $6 to $10 a pound. He had even befriended a woman at the airport who let his people stash their contraband in her truck. Each night, someone checked the truck and loaded what was needed on the appropriate flight. DeCourville bragged that he had so many illegal geoducks coming and going that even if authorities were paying attention they could never track them all. He did not buy just from Tobin. He told the informant he had "a lot of guys in on it up there."

By early fall, though, the state and federal cops had figured the system out and had settled on a plan to try and catch them all. Each time Tobin would make a deal, a cop would collect boxed clams and dump them at the airport's cargo-delivery hub. Other cops would watch one of their suspects, a truck driver from a recreational diving outfit, transfer the load and deliver it to air cargo hangars at Alaska or United. After the suspect took off, an agent would slip in and photograph the boxes to track their destinations—Fresno, Illinois, Boston, Pennsylvania. The money in Tobin's phony account started piling up, topping twenty-seven thousand dollars in just a few weeks.

Sometimes it was the rookie federal agent Al Samuels who delivered geoducks to the airport undercover. He'd spent his summer and early fall getting up to speed, trying to learn all he could about the state's clam business. He rode along with shellfish biologists on their research boat, the *Clamdestine,* and tended to the hoses as the scientists dived to survey and extract clams. He watched as they harvested the geoducks they planned to use during undercover stings. He listened as biologists explained how the legal geoduck industry was supposed to work. For the young cop it was a memorable summer. He spent most of it on the boat, baking in the sun as it glinted off the Sound's gray waters, the twin volcanic pillars of Mount Rainier and Mount Baker forming the backdrop. Samuels, by early fall, understood the Sound's illegal fishing industry and how the poachers operated.

Other times, detectives Ed Volz, Bill Jarmon, and Kevin Harrington took turns hiding at the airport and shooting video of poachers delivering stolen clams themselves. Unlike Volz and Harrington, Jarmon had spent much of his career scanning the woods and mountains for deer and bear poachers or patrolling rivers for fishermen. He'd spent far less time policing crimes on Puget Sound. But while he hadn't had as much exposure to geoduck theft, Jarmon understood the covert world. He'd once run a phony storefront for six months selling illegally caught steelhead. And he also had a feel for the intertidal zone, the oxygen-rich area where land and sea meet, the portion exposed when the tide retreats. He and his wife regularly took vacations at the shore and spent entire days digging clams off the beach.

Night after night, state detectives and federal agents packed clams by the hundreds and took them to the airport or sat in the bushes and videotaped deliveries by other poachers. Through it all an oblivious DeCourville remained confident he could outsmart mere fish cops. The agents "don't have any brains," he told Tobin. "If they had brains, they would be out there making money." If authorities confronted him, DeCourville said, he would just manipulate them. It was easier to do than most people realized, the broker added, unaware that at that moment the cops were listening in. "You've just got to make them feel powerful, any way you can," he said.

Federal agent Vicki Nomura began posing as Tobin's secretary, "Tori," so she could make phone calls directly to DeCourville herself. A few times in late September, dangerous algal blooms that can poison shellfish had closed geoduck fishing beds when DeCourville needed his clams most. On the phone, DeCourville encouraged Tori to get Tobin to collect shellfish anyway. Knowing how deadly that practice can be, the agent even tried to talk DeCourville out of it.

"I don't want [Doug] to get caught," Nomura said.

"I don't think he is going to get caught," DeCourville insisted. "There are other boats that are allowed out there doing other fishing, you know." He could just mingle with salmon boats or other legal nighttime fishermen.

"Yeah, well, I just don't want to see him get in any trouble," Nomura said.

"Well, I don't want him to get into trouble either, but I know he is a smart guy."

DeCourville had been getting divers to "pencil whip" the harvest— counterfeit documents so they could bypass health regulations. But those regulations protected people who ate clams from stores or restaurants. The rules made sure consumers didn't wind up sick or dead.

DeCourville had told Tobin that he didn't worry about poaching, but he didn't dare sell a load of shellfish without a health certificate, whether real or fake. Health officials across the country under orders from the federal government must sample commercial shellfish beds and certify that they are safe before allowing fishing. They do it to make sure that seafood isn't contaminated by pollution or toxins. The documentation then follows each shipment from the water to the restaurant. Even the most unscrupulous of DeCourville's customers wouldn't buy oysters, clams, or mussels without it. If people got sick or died from eating shellfish, anyone who'd sold it could be sued or sent to jail—unless the shipment included documents that suggested the clams were safe. "It is," DeCourville said, "the thing that clears me."

The rules are necessary because filter-feeding shellfish suck nutrients from the sea, which makes them particularly susceptible to pollution. When the shellfish draw in water, contaminants settle in their tissue. The most dangerous toxins occur naturally, are absorbed by shellfish with no warning, and can lead to paralytic shellfish poisoning in humans. The phenomenon may begin with tiny dinoflagellates, single-celled organisms that give marine waters the azure glow of bioluminescence and feed the lower rungs of the marine food web. Thousands of varieties of dinoflagellates populate marine waters. Some use sunlight to manufacture food. Others attach themselves to fish and munch on bacteria or algae. And a handful of these algae produce neurotoxins. With the right mix of nutrients, temperature changes, sunlight, or shifts in the salt content of marine water, dinoflagellates erupt and bloom like starbursts, sometimes giving waters the maroon tinge known as red tide. The blooms increase until they go dormant and settle to the ocean floor, but by then the damage already has been done. The neurotoxins get sucked up by the shellfish. In some cases, toxins can last almost a month. In butter clams the poison may stick around for two years. The neurotoxins don't harm shellfish, but they can be deadly to humans who eat them. Some of these neurotoxins are considered a thousand times more toxic than cyanide.

The symptoms can come on frighteningly fast. When shellfish poisoning strikes, the human body begins to freeze. Lips, face, and throat prickle, tingle, and go numb. Speech grows labored, and swallowing becomes impossible. Motor skills falter, and victims seem drunk. Paralysis can set in, leading to respiratory failure, coma, and sometimes death. Two years before cops started investigating DeCourville, a twenty-eight-year-old man ate several dozen raw mussels off an Alaskan beach. Ninety minutes later he was nauseated and throwing up. Within half an hour, he was on a ventilator, his pupils barely reacting to light. He lost all reflexes and did not respond to voices. Four hours later, his pupils dilated. He was essentially comatose. Then he wiggled his toes and opened his eyes. The hospital released him less than twenty-four hours later, tired but healthy. A few days later, a sixty-one-year-old woman ate fewer than a dozen mussels off another Alaskan beach. An hour later her lips went numb. Five hours after that she was dead.

The United States began testing shellfish beds in 1925, after a typhoid outbreak in Chicago, New York, and Washington, D.C., was linked to oysters contaminated by sewage. But clam smugglers don't bother waiting for official health-safety declarations. They tend to harvest in neglected regions that are more likely to be contaminated. DeCourville's suppliers often dug clams near sewage outfalls or from areas where no one had looked for harmful algae. With forged documents there was no telling if the clams were contaminated. Resulting illnesses would be untraceable, and much of the catch ended up on the far side of the world. If someone in China got sick on illegally harvested clams, no one would know whom to blame—or what part of Puget Sound the geoducks had come from, or even how many other unsafe shellfish were still out there.

It's what health officials and biologists fear most about trafficking in wildlife: globalization and the unmonitored trade in plants and animals amplifying and compounding the spread of disease. Half of the fourteen hundred most common human pathogens initially jumped from animals. New diseases, from Ebola hemorrhagic fever to West Nile virus, have been helped along by wildlife traders. The outbreak of severe acute respiratory syndrome (SARS) that later would kill hundreds in Asia ultimately would be linked to the Chinese trade in tropical ferretlike masked palm civets.

But the risk isn't just to the other side of the world. In 2003, on a farm in Wisconsin, a sick pet prairie dog bit a three-year-old girl's finger. The prairie dog later died with swollen and crusted eyes, and the child's wound blistered into a welt the size of a pumpkin seed. After developing a fever and lesions that mottled her head and face, the little girl's condition worsened until dermatologists took photographs and biopsies and sent her to a specialty clinic. Researchers eventually identified her illness as a strain of virus from the smallpox family. By then the pet vendor who sold the family its prairie dog developed lesions and chills, too.

Doctors alerted the Centers for Disease Control in Atlanta, which sent disease hunters to Illinois to investigate a shipment of coughing and sneezing prairie dogs near Chicago. Just as doctors arrived at the pet distributor's home, the CDC informed them that they'd identified the infection—monkeypox, a flulike virus found in central and western Africa that hadn't been seen in the United States in nearly thirty years. Inside the pet dealer's house they'd find out how it happened. Prairie dogs sat in a parrot cage on the floor near the refrigerator not far from pet Gambian pouched rats kept in cages on a kitchen table. Gambian rats are from Africa and are known monkeypox carriers, and prairie dogs are particularly susceptible to infections. Somehow the illness had jumped across species.

The CDC traced the rats to a shipment of African animals that included rope squirrels and brushtail porcupines brought into Dallas from Ghana. Law enforcement knew the Texas importer. Ten years earlier, he'd had an illegal shipment of Southeast Asian wildlife seized by federal agents at Los Angeles International Airport. He'd been convicted of smuggling five green tree pythons from Jakarta by strapping them around his waist beneath his clothes. He'd told customs agents he'd just been trying to keep them warm. His wildlife dealer's license had since been revoked, and he'd borrowed a customer's credentials to get the rats into the country. Many of those animals showed up dead, or dying, but survivors were passed on to pet sellers across the Midwest who sold them to families at weekend swap meets. Soon monkeypox had exploded across the heartland. Investigators eventually would see seventy-one cases. No one died, though one child at a day care developed encephalitis, and another would lose part of an eye to lesions.

Federal prosecutors eventually would wonder if the health risks of geoduck smuggling warranted stiffer penalties, but first the cops had to figure out how to make their cases. Not that the agents and state detectives could complain about their progress.

Still, some investigators remained skeptical of Doug Tobin. He was playing an intrinsic role in the investigation even though he'd gotten off to a very rough start. He'd completely blown one of his first recorded calls to an Asian geoduck buyer who had approached him with a scheme. She'd offered to overpay for clams if Tobin slipped an extra 10 percent illegally into each load. Tobin was just settling into his role and hammed it up, lacing his conversation with profanity and ethnic slurs. Investigators were aghast. They were hardly strangers to salty language—they heard it plenty when they were in the field— but Tobin's over-the-top vulgarity made the tape useless as evidence. Taped conversations got played before juries of eighty-year-old grandmothers. Agent Samuels pulled Severtson aside and suggested someone coach Tobin to tone it down. Tobin promised to clean up his mouth, but there were other troubling incidents.

Early on, after hearing the suggestion from Canfield, Tobin asked if he could sell his own legally gathered geoducks to DeCourville while he was arranging illegal undercover sales. Tobin was a fisherman, after all, and DeCourville had much of the geoduck market cornered. Tobin complained that he still had to make a living, and one sympathetic agent took his request to the U.S. attorney's office. The answer came back quickly: absolutely not. In fact, the mere suggestion made prosecutors uncomfortable. They made a special effort to remind the cops that Tobin must record every conversation with DeCourville. They didn't want to give a defense attorney any opening to imply that Tobin was making secret side deals.

Ed Volz, in particular, thought Tobin was too slick for his own good and that Severtson and Dali Borden gave him too much leeway. Volz knew Tobin whispered to his fishing buddies about working with the Feds. The informant also went off on his own to spy on suspects or engage them in conversations without recording anything. Tobin usually reported back on his encounters, but Volz considered it undisciplined. Anything incriminating would just have to be repeated on tape.

Volz was confrontational and grilled Tobin whenever he saw him, but Tobin didn't think he deserved such treatment and complained that Severtson ought to make it stop. He told the agent that Volz had "a dark air" and whined that someone who was just trying to help shouldn't have to put up with abuse. The long history of antagonism between state wildlife cops and Native American fishermen didn't help. Nor did it help that Tobin and Volz had both come of age during one of its rougher patches.

For much of the twentieth century, long after the U.S. government signed treaties promising Northwest tribes access to fish, state game wardens rousted Indian fishermen as poachers. The cops crushed boot heels into their backs, dragged them up stream banks, and tossed them in jail. When salmon really started declining in the 1960s angry white fishermen blamed Indians, even though Native Americans caught a fraction of the fish. They carved up tribal nets in the night and spit on Native competitors. Indian fishermen fought back with "fish-ins." Some drew celebrity supporters like Marlon Brando, who rowed into the Puyallup River in 1964 with tribal fishermen and pulled a steelhead from a tribal net and got arrested. "Just helping some Indian friends fish," he told reporters.

The next decade grew increasingly ugly. Wildlife officers armed with rifles lined riverbanks, and one patrol boat rammed a Nisqually fisherman's canoe. At one fish-in, wildlife agents arrived in daylight with heavy flashlights and cloth-covered clubs. Rocks, sticks, and punches were thrown. One witness, a reporter, flew to D.C. and complained about how the Indians were being treated. Another caught abusive cops on film. When clashes grew too violent to ignore, the federal government stepped in and sued Washington State, hoping to force it to recognize treaties.

Judge George Boldt's 1974 ruling shocked the region. He ruled that treaties made clear that the tribes should get an equal share of the region's fish. Suddenly the state owed half of Puget Sound's catch of fish to tribes who made up less than one percent of the population. And there were no longer as many salmon to go around, thanks to pollution, loss of habitat, and years of overfishing.

White fishermen took their anger out on Native Americans and on a new generation of wildlife patrol officers, including rookie Ed Volz. By 1976 the clashes were boiling over. A gillnet fisherman in Hood Canal rammed a state wildlife boat. One night Volz watched an onshore sniper's bullets pierce the water around him. An angry fisherman smashed Volz in the face with a dogfish shark. An officer lobbed a smoke bomb onto a gillnet boat, which set the boat ablaze. The worst came in October 1976 when a white gillnetter in a fifty-foot boat named *Alaskan Revenge* rocketed toward a small patrol boat. The officer fired his shotgun and showered a twenty-four-year-old fisherman's head and neck with pellets. One lodged in the man's brain, leaving him paralyzed.

A few years after that incident, at the end of a busy day, Ed Volz and several officers drove along Whidbey Island on the northeast side of the Sound and happened upon Doug Tobin. He was fifty miles north of his tribe's fishing grounds illegally stringing a net across a secluded beach to catch perch. Volz was exhausted and had never encountered Tobin, but figured the fisherman would work all night. So he slipped out of the car, snuck around Tobin's truck, and quietly drained the tires of air. The officers drove off and caught some sleep. The next morning a new shift arrived with a citation. They found Tobin pushing his tires uphill to a gas station.

Severtson didn't want the hassle. He agreed to separate the two men and for a while he tried to freeze Volz out. Severtson wanted to reward Tobin's enterprising spirit and saw no cause for alarm as long as Tobin reeled in poachers. Volz thought informants should be managed with caution. Severtson acted like Tobin was his pal.

Severtson tried in subtle ways to keep Tobin in check. When Tobin bragged about shooting a bull elk, Severtson quietly asked him, "As a convicted felon, what are you doing with a gun?" Tobin mumbled something about borrowing it from a friend just in time to take the shot. At Severtson's urging, Al Samuels created a fake internal memo, claiming the government had purchased an undercover minisubmarine for underwater Puget Sound surveillance. Samuels stamped the document CONFIDENTIAL. During a meeting with Tobin, Severtson let it spill from a folder so Tobin could see its contents before Severtson stuffed it back in his briefcase. The document was intended to test Tobin's loyalties. If a rumor spread among fishermen that they should watch out for government minisubs, the agents could trace it back to Tobin. They never heard a word about subs.

The agents weren't sure how to interpret that. But they couldn't argue with Tobin's results. By fall 1996, the agents and the detec-

Special Agent Richard Severtson (*with camera*) turns to Agent Andy Cohen during the arrest of a geoduck poacher in 1996. Agent Dali Borden speaks with a passenger beside the pickup. Severtson hated having his picture taken.

tives had built what they considered solid smuggling cases against DeCourville and Canfield and several other bit players. But another case working its way through the federal courts made it clear that it was risky to file charges in any case before all the others were ready to go to court; the informant's identity could leak prematurely and quickly spread before they were ready.

The cops' previous informant, Dave Ferguson, had kept a low profile since moving to Alaska to work on boats. That fall, the first geoduck-poaching arrest made with his help went before a federal judge in Washington. The judge briefly let the defendant out of jail to attend his brother's funeral in Alaska. But up north the defendant made a side trip. He took documents that showed Ferguson had been a snitch and gave them to friends who distributed them on the Alaskan docks where Ferguson worked. "The only intent in doing that was to cause some type of problem for Mr. Ferguson," a furious assistant U.S. attorney told the judge. It was a reminder that Tobin's identity could easily be revealed before the cops were ready unless the agents wrapped up every case at roughly the same time. Best to take the smugglers down all at once.

Tobin kept in touch with DeCourville to maintain appearances and checked in with the federal agents, particularly Dali Borden. When he phoned the cops with new tips, Severtson or Borden would race out the door to meet him. Borden once responded in such haste that she drove off from a gas station with the nozzle hooked in her fuel tank. Fellow agents teased her mercilessly, nicknaming her "Blaze." Tobin clearly relished the attention. He came to meetings with a yellow legal pad that contained lists of suspects. When the cops told Tobin to hide the Panasonic recorder in his breast pocket, Tobin padded the device inside a fourteen-thousand-dollar roll of bills.

Tobin seemed to know everyone and everything happening on the water. On a cloud-free spring afternoon in 1997, he met with several geoduck divers at a marina not far from the tiny hamlet of Brinnon, home of The Geoduck Tavern. He heard that one of the clam monitors was being paid off. With Agent Borden watching through the trees, he weighed out fifty milk crates of illegal clams, calling out precise numbers of pounds for the recorder.

"Where's the monitor?" Tobin asked the clam sellers.

He'll be around soon.

"What, does he get a chunk?" Tobin started counting out the money.

"About a hundred dollars," the man said but would add no more.

Two days later Tobin went to meet the same folks, narrating directions as he went, pointing out a Chevron station, traffic signs, even a garbage bin near a driveway. Inside a house he found a woman and her husband counting clams. They warned several screeching children to keep their distance. A television blared and pet parrots squawked in the background.

"So who paid the monitor?" Tobin asked again while sorting clams. "Do you want me to?"

"No, we all paid him," one diver said. They might even toss him a bonus later. "If he stops by he gets another cut."

Tobin asked if the monitor could be trusted.

They said he would be fine as long as the pressure didn't ratchet up. If it does, Tobin joked, they should just "hit him with that cash."

A week later Tobin returned to the house. A fisherman sold the informant thousands of stolen clams, then confessed he was using the profits to manufacture "bombs." Severtson later told Volz he wasn't sure what that meant, but it was something they would need to investigate.

Demand for geoducks had reached a frenzy. Prices fluctuated by the day—sometimes by the hour. Weeks earlier, the *Los Angeles Times* had described Asia's growing geoduck craze in a front-page story. The reporter traveled to Puget Sound and hung out with a diver, Henry Narte, unaware that Volz and Severtson were investigating him. The reporter watched Narte race in a speedboat to pick up geoduck samples for wealthy customers from the Pacific Rim. Narte told her he had been a crabber in Alaska but was now convinced that these clams would bring prosperity. "He is angling to make himself the middleman of choice for the Indian fishermen and wealthy Asian buyers," the reporter wrote. "'I've got the gift of gab,' he confided before racing off to shore with the clams." The reporter didn't know that most of Narte's product was obtained illegally and brokered under the table by DeCourville.

So tight was supply that buyers and brokers took ever greater risks to get clams. Casey Bakker, another broker, sold clams to a company in British Columbia, which then sold the geoducks all over Asia. In May, desperate for product, Bakker offered fishermen better money if they sold exclusively to him. For weeks, DeCourville complained to Tobin that Bakker's gambit was costing him.

Then one morning in June, DeCourville told Tobin he wanted Bakker out of his hair for good. He even knew exactly how to make it happen. DeCourville planned to hire a friend who could plant a bomb in Bakker's truck. Tobin had not bothered to record the conversation, but he immediately telephoned Special Agent Dali Borden.

chapter seven

"IT'S JUST A BUSINESS THING"

Special Agent Rich Severtson gathered his crew and outlined what Dali Borden had just heard. A few of the cops cursed their informant. Had Doug Tobin recorded all his conversations with Nick DeCourville, they would already have this on tape. Severtson hastily set a meeting with Tobin. He wanted to hear this story for himself.

Severtson and Dali headed south to a McDonald's across Interstate 5 from the Poodle Dog restaurant, where they had first met Tobin a year earlier.

DeCourville's threat wasn't entirely surprising. Wildlife trafficking attracted violence, organized crime, and other types of contraband. Severtson had seen it his entire career. Criminals saw nature as just another avenue to make money. Asian gangs and criminal syndicates from Hong Kong and Japan traded in illegal shark fins, whale meat, and tigers. Drug runners stuffed cocaine-filled condoms down the gullets of boa constrictors and packed crates of exotic snails with heroin. Federal inspectors once found a secret shelf beneath a shipment of lizards, sloths, anteaters, and kinkajous that entered Miami from Guyana. Underneath they uncovered $1 million in illegal drugs. Wildlife inspectors combing through shipments of plants and animals in Anchorage seized four hundred thousand dollars' worth of counterfeit silver dollars and one hundred fake New Jersey police badges.

The more the trade was worth the more likely it involved violence. In 1991 Brooklyn police found a thirty-nine-year-old Korean immigrant stabbed so many times he had nearly been decapitated. Blood stained the man's bedroom and living room walls, and tufts of fur littered his apartment. Jars of snakes and other creatures in formaldehyde lined the walls. A giant stuffed bear guarded one room. A collection of freezers filled another. The only things the killers took from the apartment were several dozen black bear gallbladders.

New York City Police Detective Thomas Dades worked the mystery nearly every day for five months. The victim, Haeng Gu Lee, had no job or bank account and had made his living trading animal parts, particularly bear gallbladders, which are used in traditional Chinese and Korean medicine. The gallbladders typically were dried until they wrinkled and curled like giant raisins, then they were ground into fine powder. They were used to treat cancers, burns, and liver and stomach problems. A few grams of bear bile could draw several hundred dollars. The dead man had bought from hunters who cleaved bears open near the liver, extracted the slimy fist-size pouches, and tied them closed to keep the bile intact. There were no fingerprints, and the crime was never solved, but police are certain they know what happened. Lee's customers arrived to buy his bear parts but instead ripped him off, killed him, and left the country.

Even members of Severtson's team had found themselves in harm's way. In the 1980s, Special Agent Andy Cohen saw a Canadian boat illegally hauling Dungeness crab from Puget Sound. When the boat's pilot saw Cohen, he made a beeline for the Canadian border. In a move rarely seen outside the cineplex, Cohen, gun drawn, launched himself from the patrol boat onto the crab vessel. He righted himself amid the spray just as one of the fishermen rushed toward him with a meat cleaver. Cohen disarmed him without firing a shot.

Severtson wouldn't have blood spilled over clams on his watch. But to stop it he needed to understand every nuance of the exchange between Tobin and DeCourville. Then the federal agents and detectives could map some sort of intervention. Tobin arrived twenty minutes late and repeated what the Las Vegas broker had told him: Ravenous customers were buying up Casey Bakker's clams. Desperate for a line on new supplies of shellfish, Bakker had offered Indian divers an extra two bucks per pound to sell to him. It was 40 percent more than DeCourville normally paid, and divers flocked to Bakker from DeCourville, who refused to match his competitor's price. Instead DeCourville had settled on another plan. He would hire a three-hundred-pound hit man from Los Angeles to blow up Bakker in his truck, Tobin said. DeCourville had asked Tobin to meet the guy and point out Bakker.

Severtson needed more. That night he wired electronic surveillance equipment to a nearby pay phone and told Tobin to call DeCourville back. Severtson would coach him through the call.

Without prompting, DeCourville started railing about Bakker. Earlier that day, Bakker had purchased nearly all of the available geoducks. The few that remained were more expensive than DeCourville said he could afford.

"That's not right," he barked into the phone. "I'm short yesterday because Bakker got 'em all, now I'm short today." Worse, he had big orders for Friday and Saturday, and his divers weren't working.

"I'm screwed for the week," he said. "I'm just . . . I'm dying, man."

"He's hurting me, too," Tobin said. "I'm taking it in the shorts just as bad as you are."

Tobin's nimbleness never ceased to amaze Severtson, but this time a great deal rode on his performance, including Bakker's life and Severtson's career. The veteran agent could not take chances. He needed DeCourville to be explicit, but Tobin couldn't plant ideas in his head. Nor could he push the geoduck broker where he didn't want to go. Severtson was getting antsy, and DeCourville had only briefly mentioned the "truck situation."

Finally Tobin snapped in either real or calculated frustration. He told DeCourville to spit it out. What did he want?

"I don't want to do him in," DeCourville blurted out. "I don't want to do that. I just want to hurt him."

Severtson felt some relief, but only for a moment. Just about anything could still happen.

"So, you know, how do we do that?" Tobin asked.

"Well, I can take care of the truck," DeCourville said.

What the hell did that mean? Tobin urged DeCourville to explain. DeCourville said he was at his girlfriend's house. He couldn't get into it right now. He had to be discreet.

"If that truck goes, that truck breaks down, then he's hurting," DeCourville said. "And if he's talked to, he's hurting more. You know what I mean?"

"I'm trying to," Tobin said, frustrated. "Do you mean 'if he's talked to,' you know, in a physical way?" Was DeCourville suggesting he could blow up the truck *without* Bakker in it, and then send a guy to smack him around?

"Yeah, physical."

"So is that what the three-hundred-pounder is for?"

"Yeah," DeCourville said of the hit man he planned to hire. He snorted. "He's a sheriff, retired."

"He's a retired sheriff—that's great," Tobin said.

"The boys use him all the time," DeCourville said.

"Well . . . it sounds like you've mellowed since this morning here," Tobin said. The scenario, while serious, was not as straightforward as Tobin had described to Severtson. Was DeCourville having second thoughts?

DeCourville changed the subject. They talked about Tobin's family, briefly, and chatted about DeCourville perhaps going to Canada to get geoducks.

Then DeCourville said, "I'd like to see an arm or leg, you know? You know what I'm talking about?"

Tobin did. "I understand. Arm and a leg will work," Tobin said. "So . . . what can I do to help you? I'm trying to relieve the strain on you. Know why? Because I need you to buy geoducks. So the less strain you have the longer you're going to fucking live. That's a Native thing, Nick."

"The man's name is Rick," DeCourville blurted out. "About three hundred pounds, blond hair. All I want you to do is point him in the right direction . . . I don't want the wrong guy hurt."

Once Tobin was off the phone, the agents talked about this. Rick. Shit. No last name. They still didn't know anywhere near enough.

The next day Severtson and Special Agent Al Samuels drove south to meet with Casey Bakker.

Bakker, a fast-talking broker with a quick wit and disarming grin, had been a fixture in the geoduck industry since 1980. His father had worked for an oil company, and Bakker as a kid had moved every few years, the family landing in many of the country's energy hot spots: Denver, Houston, Pittsburgh, New Orleans. He'd come to Washington to attend Evergreen State College in 1976 but ran out of money toward the end of his senior year. He'd learned scuba as a kid on Colorado's high lakes. The next thing he knew, he'd stumbled into the industry that gave his alma mater its mascot. Seventeen years later—with a house and a family and a clam buyer's license—Bakker found himself face-to-face with federal agents.

Severtson, Samuels, and Bakker stood outside a Starbucks in Tacoma. The agents warned the clam salesman that he might be the subject of a contract hit. They offered protection, but Bakker was prickly. He asked if some wiseguy was going to jump out of a black Lincoln with a crowbar. Severtson's answer ignored the sarcasm. Yes, that's exactly what could go down, Severtson said.

Bakker found the cops' concern disconcerting and wasn't sure what to make of it. He knew DeCourville only as a businessman and competitor. But he believed that under the right conditions people were capable of just about anything. Bakker didn't like being bullied—not by DeCourville and not by the cops. He told the agents he'd take care of himself. Severtson said that was the very thing that concerned law enforcement. He begged Bakker not to do anything rash. Bakker said he'd be fine. He would move underground. Severtson relented, recognizing a lost cause, but quietly planned to send someone to keep watch on Bakker anyway.

The agents weren't clear on just what DeCourville wanted. They couldn't be sure he wouldn't try to kill Bakker. It was possible DeCourville just didn't want to tell Tobin, so they cooked up an excuse for their informant to call back and ask for clarity.

This time DeCourville didn't equivocate. He had contacted his helpers that morning. Casey Bakker, he said, "had to have the shit beat out of him, plain and simple."

"They're going to call me back tonight with the cost and make arrangements. They're going to use one guy from up there and another guy from L.A."

"Wow," Tobin said. "They're that good, huh?"

"Let me tell you something in confidence," DeCourville said. "I used to do things for the mob. These are mob guys, OK, and this is a retired sheriff who used to work for the mob, while he was a sheriff . . . a three-hundred-pounder. These guys are pros. He trained Robert De

Niro in a movie, you know, about how to be tough. He's expensive, but see, I got a friend. We have mutual friends."

Tobin asked to hear more.

DeCourville sounded happy to be encouraged. "I got to tell you. Got a minute?" He settled in. "I had a guy cheat me on twenty-seven thousand dollars. Oh, this was back ten years ago. I called a guy in and he went over to his house at night, broke in, and was standing there by the bed so silently nobody knew it. Plugged a [defibrillator] in, put a piece of tape over his mouth—that woke him up—and says, 'This is the first and last visit. When I put this on you and turn the juice on, you're going to have a heart attack. I'll pull the tape off, won't be a mark on your body, and you'll be in the hospital.'" His friend applied the paddles, DeCourville said, and shocked the guy, who convulsed and went into cardiac arrest. Then his friend called 911. The ambulance came and took the guy to the hospital.

"I'm up there the next day with some roses," DeCourville said. "I said, 'I don't want to bring you roses twice, OK?' I said, 'The next time it will be on your bed. Your deathbed.' And, ah, I got my money. When he got out of the hospital, I got my money. Bang. I don't know what the hell he did to get it, but he got it."

Tobin pushed on with Dali and Samuels at his side. "This guy in Seattle . . . is he an ex-cop, too?" Tobin asked.

"No," DeCourville said. "He's part of the fucking Mafia."

"So I don't know him, right?" Tobin asked.

"No," DeCourville said. "He's a Mafia guy."

Tobin told DeCourville to be sure and let him know when the deed went down. Tobin told the broker that he wanted to be two hundred miles out to sea with friends when it happened so he could be free from scrutiny.

"I feel, as a friend, I feel, you know, pretty safe saying that, that you're my friend," Tobin continued. "You've been there and I hope I've been there for you."

"You've been there for me. I would . . . I treat you like a brother," DeCourville said. "I treat you just like I do my brother."

"So we've got a common interest here," Tobin said. "And the sooner this cocksucker quits gouging into my pocket and yours then the better our relationship . . . I mean, it's just a . . . business thing."

DeCourville said he didn't know precisely when things would happen, but he had been told the attack would take place in fewer than ten days.

"These guys don't miss," DeCourville said. "They do precisely what you tell them to do . . . They're professionals and they'll disappear. These guys are pros. They don't fuck around."

"Well, look . . . there's no way this is going to come back on you, is it?" Tobin asked.

"I don't see how," DeCourville said. "You're the only link and I trust you."

Tobin clarified. He was talking about Bakker. Wouldn't it be obvious to Bakker that DeCourville was responsible?

"He knows you're connected," Tobin said. "Who else can sit in the middle of a fucking pile of sand and control eighty-plus percent of all geoduck sold in the U.S. . . . and never touch one of the son of a bitches? It's a no-brainer. The guy knows you can pick up the phone and have him taken out."

"I don't worry about that," DeCourville said. "There's no way you can trace what I'm doing. It's clean."

"Because I need you to buy geoduck," Tobin said. "That's honesty."

"They give me a mobile phone in jail and I'll still sell ya geoduck," DeCourville said. Then he added, laughing, "I'll still never *see* the sons of bitches."

They still didn't know who the hit man was, but Severtson would make the other agents track him down. They had to. Severtson would order special agents Andy Cohen and Al Samuels to trace all the calls DeCourville had made or received in the last three days. In the meantime the two cops would work the phones. They knew the man DeCourville mentioned was an ex-sheriff's deputy, probably in Los Angeles. They knew his weight but had only a first name: Rick. The agents needed to reach this guy before he reached Casey Bakker. Severtson needed to alert the U.S. attorney's office. Severtson told his people: Find this guy now!

Assistant U.S. Attorney Helen "Micki" Brunner was incredulous. "Are you kidding me?" she asked Severtson. "You've *got* to be kidding me."

Severtson was in Brunner's downtown Seattle office. He'd just told her and Detective Ed Volz about DeCourville. Brunner and Severtson had known each other for years. In fact, Brunner knew most of the state and federal wildlife agents and was well versed in the geoduck investigation. She had been involved for months and would prosecute the cases they brought forward. She knew her way around wildlife crimes. But a hit man? Over clams?

In Brunner the agents had an enthusiastic champion. She was an outdoorswoman, a hiker, an unstoppable skier, and a kayaker who liked watching ospreys dive in the bird-rich Nisqually River Delta. She had handled cases for the Environmental Protection Agency and had come to Seattle from the Environment and Natural Resources Division of the Justice Department in Washington, D.C.

Brunner brought an encyclopedic knowledge of environmental law. She had taken on timber companies, metal-plating operations, and plastics manufacturers. Once, she had sent steel-company executives to prison for hiring a driver who dumped hundreds of drums of toxic sludge in a Washington cow pasture. She went after a city official who ordered hazardous waste buried in a sandy trench, where it leaked into the Pacific Ocean next to a wildlife refuge.

Brunner had become the go-to prosecutor on wildlife-trafficking cases. She was smart and decisive and had no trouble recognizing the urgency after Severtson filled her in about Casey Bakker. They needed to resolve this before someone got hurt. She told Severtson to let her know what else he needed.

Severtson checked in on agents Cohen and Samuels as they made calls. Samuels, the computer whiz, was told to work his digital magic on law-enforcement databases. If the hit man had been a sheriff's deputy, maybe they could find him that way. The two agents scrambled for several hours. Severtson repeatedly stuck his head in their cubicles and demanded updates.

By morning, the agents were certain they'd found their man: Richard "Ricky" Jones. Jones had worked as a Los Angeles County sheriff's deputy during the 1980s. He retired on a partial disability after being struck in the head by a baton during training. Since leaving law enforcement he had run a security company with a Beverly Hills Police sergeant, worked as a bodyguard, and been a driver for convicted mob figures. The agents got a copy of Rick Jones's driver's license, which showed a picture of a baby-faced six-foot-two man with a wide forehead and jowly neck. The license listed his weight at 280 pounds. It had to be him.

On a Sunday in June 1997, Jones rolled out of the Los Angeles basin and over the San Gabriel Mountains and cut across the Mojave to Las Vegas. In his younger days, he'd been a bouncer and kickboxer who bragged that he'd fought prizefights in Mexico. He was headed to Vegas to work as a consultant for a friend who published personal security newsletters with titles such as *Weapons & Tactics for Personal*

Defense. For fifteen dollars, the company also mailed gamblers lottery picks from a secret strike-it-rich system they claimed had been passed down from Al Capone.

Jones had made the drive often in recent months to proofread security articles. The two men also were making a video: *How to Recognize and Avoid Violent Street Crime*. Jones checked into the Heritage Inn Best Western and, for the next two days, hopped down to the Tropicana to check out a survival seminar.

Wednesday morning Jones's friend paid him five hundred dollars for his work and gave him a new Colt Python revolver. Jones checked in with an attorney in Los Angeles who told him an acquaintance, Nichols DeCourville, wanted to see him about some investigative work. In the 1980s, the attorney's law practice had been across the street from Nick's Fish Market.

Jones pulled up around 1 P.M. DeCourville lived five miles down Tropicana Avenue from the Strip, near the end of a short cul-de-sac. His two-story white stucco town house had a red-tile roof and a black wrought-iron gate across the front walkway. His girlfriend lived two doors down. DeCourville was bare-chested and wore a ponytail. The maid was cleaning inside, so he led Jones to the backyard. Fifteen minutes later, Jones was headed back to Los Angeles.

Severtson had an idea. Tobin should convince DeCourville to hire *Tobin* to do the job, instead of Jones. That was Severtson's way. Take what circumstances provided and turn them to his advantage.

On June 9, DeCourville and Tobin were back in touch.

"How are we coming with our little buddy?" Tobin asked.

"Ahh, well, I don't want to talk about it," DeCourville said. "It's coming—it'll happen."

Tobin backed off but tried again that evening, calling from a pay phone at a truck stop along Interstate 5 outside Fife. Beside him, agents Al Samuels and Dali Borden listened in.

"How's it going tonight, OK?" DeCourville asked.

"No," Tobin told him. "It's not so good, my friend."

Bakker had raised the price yet again. Less than two weeks ago, DeCourville had paid $4.50 a pound for geoducks. Now Bakker was offering divers $8 a pound. DeCourville couldn't afford to buy at that price.

"That puts me out of business," DeCourville said. "Fuck. Couldn't get any, huh?"

"I couldn't at the price you wanted to pay," Tobin told him.

"Jesus Christ," DeCourville said. "Goddamn. I'm in trouble."

Tobin offered to help. Tobin said he had a few tough guys coming over from the coast to help him settle an old score. If DeCourville was inclined, they could take over the Bakker job. "Without a fucking question, they can take care of this little problem you want done here," Tobin said.

His answer came quickly.

"OK, I'll call off the other," DeCourville said. "I'm losing, you know—I'm losing my ass."

Tobin agreed. But, he asked, what exactly did they want done to Bakker?

"I'd say punch him in the nose, break his fucking arm, and tell him to stay away from the Indians in the south Sound, period," DeCourville said. "Don't you think that's enough? I think that's enough. And if he doesn't listen . . ."

Tobin told him the job would cost five thousand dollars. DeCourville agreed to pay, after some grumbling. He had easily lost that much in the last week, or even in the last twenty-four hours, DeCourville said. If Tobin faxed DeCourville a fake invoice, DeCourville would wire a portion of the money the next day and cover it through his business. The rest he would send later in the week. He didn't want to send a lump sum: "That'll show out like a sore thumb."

Samuels prepared a fake bill for the purchase of 546 pounds of geoduck, valued at $3,003. He backdated it to January 5, and stamped it with a seal from Tobin's company, Blue Raven. At the bottom, Samuels wrote: CORRECTED INVOICE FROM MISBILLING. Tobin signed it. Samuels paid $2.50 to the clerk behind the counter at the truck stop, who faxed it to DeCourville.

Three minutes later, Tobin called DeCourville, who suddenly seemed eager. "I got to get this thing done. Is it, is it being handled? As we speak?"

Tobin said he'd already called the guys. It could happen as soon as the next day.

"Good," DeCourville said.

Tobin joked that when it was over, DeCourville should send the boys from Las Vegas up to Seattle for lessons on how to really get things done. Then he reminded DeCourville what he had paid for: a broken nose, a broken arm.

"Leg would be better," DeCourville corrected, then muttered, "That son of a bitch."

"Everything's a done deal. Everything's in motion," Tobin reminded him. "I didn't know if it was an arm or a leg so I gotta make one more call and yeah, it'll be a leg."

"Just tell 'em it'll be an arm, the next time it'll be a leg or vice versa," DeCourville said. Then, laughing, he added: "Just make sure he's not to go near the Indians in the south Sound. That's the message."

"I made that abundantly clear," Tobin said. "They'll pound that into him."

Three days later, Severtson and Borden drove south from Seattle again, crossed the Tacoma Narrows Bridge to the Key Peninsula and visited Seafirst Bank, where they'd opened Tobin's business account. The wire transfer had arrived from Nevada.

On Friday the thirteenth, Severtson and Borden met again with

Tobin at their office in Seattle. They called DeCourville, catching him just after 10 A.M.

"How is everything with the headache problem?" DeCourville asked. "How is that coming along?"

Tobin said he'd paid for half upfront. The rest would come when the job was finished. DeCourville told Tobin to send a new invoice, this one marked for the purchase of an ice machine. The next day, $1,997 arrived in Tobin's account.

By Monday night it was over. Tobin called DeCourville one last time around 10:30 P.M.

"You taken care of business?" DeCourville asked.

"I'm here to report that your money was well spent," Tobin said. Tobin actually sounded almost joyful.

"Oh, that's good," DeCourville said. He started laughing and made a joke.

"Nick, listen to me," Tobin interrupted. "Nick, I kind of think they went just a little overboard . . ."

"They didn't throw any names around did they?" DeCourville asked.

"Absolutely not," Tobin said. "They just threw him around."

Tobin told DeCourville he had to meet the thugs the next day to finish paying them their share. DeCourville said they should lay low. But within two weeks, he wanted Tobin out getting more geoducks for him.

"I'm hard up for tomorrow, I'll tell you," DeCourville said.

"I know you're hard up, Nick," Tobin said.

"Next week, all this week and next week, and then it will be balls out," DeCourville said.

"Balls to the walls, huh?" Tobin asked. "All right."

A phalanx of state and federal agents descended on Las Vegas after midnight and settled in for a short, restless night at the Excalibur. An hour after dawn the next day, June 19, 1997, Rich Severtson, Ed Volz, and a line of other officers crouched around the corner from DeCourville's home off the Strip. Vicki Nomura, posing again as Tori, Tobin's secretary, walked up alone and rang the bell. No one answered, so she called DeCourville's cell. She said she was in Vegas and needed to see him. DeCourville told her he was at his girlfriend's house down the street and would be right there.

Days earlier, another Washington State wildlife investigator, Ron Peregrin, had bought wraparound casts, arm slings, and Ace bandages and had driven to Casey Bakker's Olympia-area home, which was dark and shuttered, as if abandoned. Peregrin had found Bakker inside, had bandaged him up, and put him in leg braces and an arm sling in preparation for a photograph to show around the docks. The cops had hoped DeCourville would get word of Tobin's hit from other fishermen.

DeCourville stepped outside a few doors down. Nomura waved—a signal to the agents to move in. DeCourville wandered up to his front door. The broker shoved his hand into his pocket for his keys, and Severtson and Volz pushed him against the wall. They patted him down for weapons and led him inside.

The agents fanned out and searched each room. Like the cherry 1984 Lincoln in the garage with fewer than twenty thousand miles on it, DeCourville's home was elegant and immaculate. Indian art and Buddha statues adorned the shelves and halls. Downstairs, the agents found a tidy computer room with monitors linked to outdoor surveillance cameras. DeCourville had run cables from his office out his window and down the street, linking to a room in his girlfriend's home where he could work. Upstairs, searchers found little of consequence: copy machines, file cabinets, a rolltop desk, a stereo. They seized a tiny, loaded .25-caliber Beretta from under a bed.

Ed Volz prepares for the search outside Nichols DeCourville's home in Las Vegas.

A computer expert tackled the hard drive while Volz and others went through the house. Severtson sat before a glass dining room table and started in on DeCourville. He moved cautiously at first. He had invited the FBI to participate in the interview, but the agents were late. They'd been in a car accident on the way to DeCourville's house. Severtson read DeCourville his rights and asked about the suspect's business, his personal life, and about seafood.

DeCourville seemed at ease and eager to oblige. He smoked his pipe, popped vitamins, and told the investigators how he planned to stay fit during his golden years. He told the agents he never bought illegal products, nor had he even seen a geoduck until the previous month. He made it a point never to touch the things.

When the FBI arrived, Severtson shifted gears and mentioned Casey Bakker. DeCourville blinked violently and his face twitched.

He repeatedly asked for water. DeCourville's demeanor changed so abruptly that several agents remarked on the shift in their notes. Severtson asked what DeCourville had done to Bakker.

DeCourville furiously smoked his pipe. He drew so much on his tobacco that a growing haze of smoke hovered above his head. He said he'd heard something about Bakker getting hurt but didn't know anything about it. After Severtson said the agents didn't believe him, DeCourville insisted he would never physically harm a competitor. That would be dumb, and he was not dumb. "Not my style," DeCourville said.

Severtson showed DeCourville a picture of Rick Jones. DeCourville said he'd never seen the man. Severtson told DeCourville that agents *knew* Jones had been in the house. DeCourville said he suddenly remembered.

So it went. Asked about the wire transfers, DeCourville pretended he owed Tobin for refrigeration work. Told his phone calls had been recorded, DeCourville's ticks accelerated. "I don't want . . . If you've got things on tape, I don't want to . . ." He did not finish the thought. "What do you want?" he finally asked.

Severtson told DeCourville that the cops wanted the truth.

At 6:35 A.M., Detective Kevin Harrington and several other officers moved toward a two-story blue-and-white farmhouse in Olalla, Washington, across from Puget Sound's Vashon Island.

They crossed the damp yard and saw their subject, Gene Canfield, sipping coffee on his porch. A federal agent with Harrington handed Canfield a copy of a search warrant, and Harrington pulled him aside. While a team searched the house and gathered sales records, Harrington told Canfield he was not under arrest, but that could change.

He encouraged Canfield to share information. Canfield laughed. "No offense," the fisherman said, but "I'm a little distrustful."

The men chatted, and Canfield loosened up. He acknowledged working with fishermen he knew were poachers. He even joked about laughs he'd shared at investigators' expense. He told Harrington about a day when he and a group of men sat at a diver's house, sipped beer, and watched police surveillance videos of their host loading stolen clams onto a truck. Harrington was annoyed. One of his colleagues had shot that video two years earlier for a case that prosecutors never took all the way to trial. The tape must have been turned over to the suspect's defense attorney during discovery.

The agents in the house disconnected Canfield's computer, seized the hard drive, and wrote down phone numbers from his caller ID. Harrington urged Canfield to reconsider. He could help himself by cooperating.

"I know how the game works," Canfield said. He would get a public defender and "get on board."

That same morning, a caravan of vehicles led by a Washington State Patrol SWAT team sped toward another cul-de-sac, this one along the southern edge of Puget Sound's Hood Canal. The wildlife agents knew this fisherman—the one with the parrots and screaming children who told Tobin he bribed official geoduck monitors—kept weapons in the house. They feared what might happen if they didn't catch him by surprise. The SWAT team, in full black tactical gear and carrying stun grenades, hit the house hard and fast. The cops battered down the door and moved room to room, clearing family members and rousting the geoduck diver.

They dragged the fisherman to the kitchen, where agents interviewed him. The fisherman kept laughing and staring off at his pet parrots. Searchers hunted for evidence of geoduck poaching, but after a few minutes one shouted from upstairs: "Ron, you need to see this!" An upstairs bedroom was crammed chest-high with antique furniture, lamps, chairs stacked on end tables, and an unusual assortment of cardboard shoe boxes. Inside the boxes, Detective Ron Peregrin found drilled-out tennis balls, cylinders cut the length of toilet-paper tubes, and hundreds of fuses. Scattered about the room were a number of coffee cans filled to the brim with some type of gunpowder. The wildlife cops called the SWAT team's explosives expert, who called in the bomb squad.

The fisherman had been using his geoduck profits to buy raw materials to make bombs to sell as fireworks. Apparently not satisfied with homemade M-80s, he stuffed tennis balls and fist-size canisters with gunpowder to form grenades. A misplaced match would have leveled the house and taken out a good portion of the neighborhood. The bomb squad seized enough explosives to fill a truck.

Also that day, two FBI agents met Rick Jones at his Los Angeles apartment. They waited patiently while the ex–sheriff's deputy called his employer to say he'd be late. Jones told the agents he did security and had worked as a driver, hauling around comedians Sam Kinison and Rodney Dangerfield. He said he occasionally drove for Dominic Montemarano, aka "Donny Shacks," a capo in the Colombo crime family. Jones talked about his street-survival video and told the men about meeting DeCourville.

Jones said DeCourville had complained about a guy in Seattle undercutting his prices. DeCourville never mentioned violence, but his hand gestures left no mistaking his meaning: DeCourville had wanted to hire Jones to hurt this man. Jones said he knew DeCourville would find a high sum outrageous, so he suggested a flat fee of fifteen thousand dollars. DeCourville kicked him out. Jones told the agents he never would have gone through with it. He said it wasn't part of his job description to physically harm anyone and that he'd only wanted to "fuck with" DeCourville. Before they left, the FBI agents hit Jones with a subpoena.

Two weeks after the raids, Special Agent Andy Cohen, with the National Marine Fisheries Service, badged his way through security at Sea-Tac International Airport to pick up Jones for his appearance before the grand jury. Cohen took a seat on a bench. The agent didn't know whether anyone might have flown with Jones, or whether some criminal type might try to contact him between the gate and the car. The agent didn't know what he was dealing with; someone could pass Jones instructions or cash to try and influence his testimony, so Cohen hoped to observe his witness unnoticed for a few minutes. Severtson constantly pounded his agents to practice their clandestine skills. To Cohen, a James Bond devotee, this seemed like the ideal moment. He poked a pencil-size hole in a newspaper and pretended to read while keeping an eye on the gate.

There was no mistaking Jones. He stepped off at 10 A.M., the last person aboard. He was a tank, with a fleshy chin that drooped diagonally to his neck. In early July, high summer, Jones wore a turtleneck beneath a sport coat. He planned to return to Southern California that night and carried no bags. He carried only a child's stuffed animal.

Cohen made his introduction after watching for a few minutes. He asked Jones if he had ever been to Seattle. Did he know anyone here? Jones assured him he did not, but Cohen wouldn't let it go. He was watching Jones's face to see if he was lying. Are you meeting someone? Jones again said no.

"I figured maybe the bear was a gift?" Cohen finally asked.

"This," Jones said, "is *my* bear."

At the Federal Courthouse in downtown Seattle, Cohen walked Jones to a waiting room and asked a series of questions, but Jones gave Cohen the same story he had told the FBI two weeks earlier. After forty-five minutes with the grand jury, Cohen drove Jones back to the airport. Cohen couldn't stand it any longer and blurted out: "What in the world is up with the bear?"

Jones told him his first wife had died of cancer. He missed her so badly he was unable to sleep. At night he curled up with the stuffed animal on the floor. The ex-deputy DeCourville had tried to hire as a hit man said he was afraid of flying and took the bear everywhere. It was his security blanket.

chapter eight

AN INCREDIBLE VIRUS

Flipping through the DeCourville files, Assistant U.S. Attorney Micki Brunner saw what she would need to bring the loose threads together in court. The case had great potential, and Brunner knew what such cases required. Brunner had cemented her reputation five years earlier. She, Agent Andy Cohen, and two state wildlife detectives had brought down thieves who illegally killed sturgeon from the Columbia River and ripped out their eggs to sell to a New Jersey man *People* magazine had dubbed "The King of Caviar."

Sturgeon, monstrous and bony fish that lived alongside and outlasted the dinosaurs, are among the world's oldest and largest freshwater creatures. They can reach twenty feet, live a century, and weigh more than half a ton. Aside from salt, their eggs are the sole ingredient in caviar, which fetches higher prices than some precious metals. But by the mid-1980s, declining numbers of Caspian Sea beluga sturgeon and limited access after the Iranian revolution made it difficult for U.S. suppliers to get their hands on high-grade caviar. Producers turned to other sources, including white sturgeon from the Columbia, a river that marks the border between Oregon and Washington. But so few of the biggest slow-growing fish work their way along the river bottom that both states strictly limit how many can be caught.

Stephen Darnell knew just how to snag these huge fish, but he seemed an unlikely candidate to earn prison time for poaching them. Darnell was a great fisherman who lived near the base of the Grand Coulee Dam in central Washington. So fond was he of the Pacific Northwest's sturgeon that he'd once written a letter to his local newspaper appealing to citizens to help save them. But Darnell was also an inventor who was low on funds and in 1985 he jettisoned his save-the-sturgeon principles for a fillet knife and piles of cash. His crimes might have gone undetected even longer if two thieves hadn't finally robbed a small bank near Vancouver, Washington, in 1990.

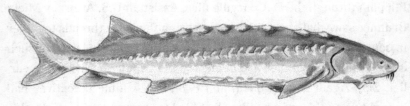

White sturgeon, *Acipenser transmontanus*

FBI agents tracked the cash stained by an exploding red dye pack to the Value Motel, where a helpful desk clerk told the agents about the strange guests in room 124. Two men had paid $968.40 in advance for a month's rent and requested no maid service. Another motel employee noticed a stench escaping from the room, though, and let himself inside. He found boxes that held white plastic containers with screw tops, along with fishing poles, waders, two-quart jars, a calculator, an outboard motor, and buckets of liquid that smelled like fish. He presumed he'd stumbled on a meth lab.

The FBI agents hid for several days in a neighboring suite and watched room 124 through the blinds. They saw a man dump an empty container of salt into a trash bin and two men going to Federal Express to ship hefty packages to a caviar company in New Jersey. The agents eventually realized these men hadn't robbed a bank, and called in wildlife cops. State Detective Paul Buerger and Agent Cohen tracked shipping and bank records to find that the New Jersey buyer was a company run by Arnold Hansen-Sturm, who came from a long and storied line of caviar distributors. He supplied the finest restaurants and hotels from Europe to Japan, his top prices sometimes hitting $750 for a fourteen-ounce tin.

Yet caviar is actually quite simple to create. Catch a whiskered bottom-dwelling sturgeon, gut it, and remove its egg sac. Place the gooey mess over mesh and separate from it the mass of eggs, called roe. Rinse the roe, mix the eggs with salt, pack them in tins or squeeze them in cheesecloth, and process them into paste. It's a decidedly low-tech luxury item, requiring little more than running water and a plastic bucket.

Darnell was resourceful enough to realize that a cheap motel room would suffice. He and a partner began catching and gutting Columbia sturgeon and shipping their product to Arnold Hansen-Sturm. Over five years, Darnell mailed nearly seventy packages of caviar, more than thirty-two hundred pounds. For the men to have collected that much roe, biologist would later calculate that they had killed at least two thousand of the biggest adult fish—almost 10 percent of the largest sturgeon in the Columbia River.

Wildlife cops often faced trouble convincing lawyers their cases mattered, and they tended to grumble most about federal prosecutors. Because the Justice Department likes victories, prosecutors often let defendants plea-bargain their way to lesser charges, which in wildlife cases can result in little or no punishment. When presented with

Darnell and Hansen-Sturm, Brunner needed no arm-twisting. The caviar baron had paid the fishermen $247,000 and sold the low-grade product to his least-discerning customers, mislabeling it as imported beluga and earning well over $1 million. Brunner insisted that all three men serve prison time. The case received national attention and helped spark a broader look at the American caviar trade. It also earned Brunner the lasting respect of wildlife cops.

The DeCourville case could also make headlines. The geoduck kingpin had sold hundreds of thousands of dollars' worth of illegal shellfish, and agents had intercepted and prevented a violent crime. Between the paperwork, the recorded calls, and the wire transfers, the detectives and agents really had him nailed. There were the cases against Gene Canfield and several other middlemen, too. Severtson and Dali Borden had used Tobin to document several illegal transactions. They also had the bomb maker's case. Making it all stick would come down to Micki Brunner, but she saw a shadow hovering over everything.

Tobin was everywhere in this case—on the phone with DeCourville and Canfield, face-to-face with the bomb maker, and in contact with all the other minor players. Brunner needed Tobin to testify, but she could tell he might prove less than ideal as a witness. No one expected informants to be saints, especially prosecutors, but Brunner was worried. Tobin had been in and out of trouble his whole life, with a criminal record that included serious crimes, such as manslaughter. She had never met Tobin, but she would need to meet him soon. The way he described his past would dictate how she proceeded.

Tobin's record was substantial. He had been sent to the state penitentiary twice. He argued both convictions were mistakes and had convinced many people he was right—he seemed too gregarious to be in any way a dangerous criminal. He was first incarcerated at twenty-one in connection with the assault of a former all-American defensive tackle at Washington State University. One day, a decade past his prime, boredom, testosterone, and perhaps a spot of depression had led the athlete to arm wrestle Doug's older brother John. John Tobin was a powerful young man, and soon the arm wrestling escalated to a boxing bout. Witnesses claimed the football player won, and a few days later, on a cold rainy September evening, someone entered the ex-WSU star's apartment and beat the man senseless while he slept in his bed.

Police arrested John Tobin. Prosecutors convicted him of assault and sent him to prison. After the trial, Doug came forward saying he had new evidence. John got a new hearing. Doug said he had driven another man to the athlete's house that night and waited in the car while the man went into the player's apartment to collect a debt. Doug told the court that he was still sitting there when the man sprinted out, shouting, "I got him!" The judge was unimpressed and denied John a new trial, but the prosecutor seized on Doug's words. He'd just confessed that he drove a getaway car for *someone*. The prosecutor charged and convicted Doug as an accomplice, and he was sent to the Washington State Penitentiary in Walla Walla.

Doug's attorney insisted his client had been railroaded. A columnist for the *Seattle Times* in 1977 argued Doug's conviction had been a misapplication of justice, a travesty: "The things that happened in this case," the columnist wrote, "can knock you over." Doug ended up serving three years, longer even than his brother. After his release he moved to Alaska and worked as a logger but returned to fishing Washington's waters after his father died in 1982.

Four years later, Doug was facing a murder charge. Joanne Jirovec was forty-eight and working as an administrative aide at Olympia's Evergreen State College. On March 22, 1986, her husband, David

Jirovec, called from a suburban Denny's to say the family motor home kept stalling. He asked her to run his tools to the restaurant. By the time Joanne arrived in her van the other vehicle was running fine, so the couple went inside, drank coffee, then left for home in separate vehicles.

The next day, police found the Jirovec family van abandoned beside a freeway outside Chehalis, a farm community dotted with dairies and a mint plant. Joanne's body was curled in the back. She had been shot three times in the head. The van's CB and radar detector, along with Joanne's purse, were missing.

The execution of a college worker attracted a lot of attention. A thousand people showed up for Joanne's funeral. Her husband seemed inconsolable. But over several months, detectives pieced together an unseemly tale. The Jirovecs had been quarreling and faced serious marital troubles—David had had at least three affairs. It was no secret that David wanted his wife gone. Witnesses had heard David repeatedly ask an ex-con fisherman if he knew anyone who would kill his wife.

Tobin told police he had dismissed Jirovec's solicitations three times as the requests of a loon. When Jirovec showed Doug ten thousand dollars he kept stashed in the Dodge van, Tobin said he gossiped about it to an out-of-work friend named Daryl Burns. Tobin and Burns talked about how easy it would be to take the money. Tobin told police that his friend was just supposed to steal the cash on the night Joanne Jirovec disappeared. Doug said he had no idea that Burns would kill her.

Tobin cooperated with police. Burns did not, until it became clear he was going away. He confessed to being the trigger man, but claimed Tobin had said Burns would be paid. Burns and Jirovec were convicted of first-degree murder and sentenced to life in prison. Tobin pleaded guilty to manslaughter. He maintained all along he thought he'd aided a botched robbery, not a contract killing.

In prison Burns backed Tobin's story. He recanted his confession and insisted he'd made it under duress. At another point, he said a fourth party had done the shooting.

Tobin's story sounded implausible, especially to the fish cops, and Tobin knew it. But to the detectives Tobin's past didn't matter all that much. They cared more about his ability to gather facts—and the truthfulness of the information he provided. Even the most forgiving among them, Agent Severtson and Detective Harrington, assumed Tobin actually was guilty. Severtson had been a highway patrolman in Oregon, and Harrington had worked in a juvenile-detention center around teenage murderers. Both had spent so much time around thieves and killers that they didn't waste energy analyzing reputations. Besides, many of law enforcement's most famous informants either witnessed or participated in dozens of murders. The cops were interested in how their informant performed.

But Brunner needed to think strategically for the courtroom. Before an informant testified she usually walked him through his life. Brunner needed to hear what he said about his past and how he sounded when he said it. She needed to understand whether defense attorneys could use him to submarine her case. A prosecutor needed to know if her star witness had problems with the truth. Brunner could not call a liar to the stand.

Three weeks after DeCourville's arrest, Doug Tobin showed up at Brunner's office. By then she knew his background well. She also knew her prosecution faced serious trouble. Brunner's position was clear: Tobin would have to admit his involvement in the Jirovec and assault cases. He might still be credible if he admitted his role. If he did not, defense attorneys would shred him.

Seated at an oak table in a spacious conference room with agents Severtson and Borden, Brunner asked Tobin to tell her about his history. As Tobin spoke, Brunner pressed with increasing vigor. Why would Jirovec feel he could approach Tobin about a murder—and more than once? Why didn't Tobin call police? Why did he tell Daryl Burns about the money? Why did he plead guilty if he didn't do anything?

Tobin was a physical being. He often employed his body as an instrument of enthusiasm or intimidation. He'd touch an elbow in a playful gesture to sell a point, or lean close and lock eyes and glower like a caged bear. Tobin found Brunner vexing. He towered over her and outweighed her by 150 pounds. But Brunner was unfazed by his indignation. Over and over Tobin claimed he had done no wrong, and still Brunner grilled him fiercely. She was surprised that Tobin seemed so unprepared for confrontation—he had been in and out of court enough. Tobin barked and huffed and stuck to his story no matter how unlikely it sounded. At times, Agent Borden even jumped in to defend him.

By the time the meeting ended, everyone was on edge. Brunner felt her time had been wasted and that she had been lied to repeatedly. Tobin later complained about being dismissed by the cops and lawyers he had worked so hard to help. For Severtson and Dali Borden, the consequences were worse: Every case they had made with Tobin as an informant was now tainted. If they wanted to convict and put away poachers and smugglers, someone would have to reinvestigate their cases.

No one stated the obvious: It appeared a year's worth of work had just spiraled down the drain.

Ed Volz tried to keep silent. He had always feared it would come apart. Now, thanks to Tobin, it had.

Outwardly, Severtson took the meltdown in stride, but Volz knew it was a tremendous blow. Severtson's own agent, Dali Borden, had too quickly and passionately defended Tobin. Borden had gotten worked up in the U.S. attorney's office. Severtson felt she'd grown too attached. Tobin had somehow spun her head around, a testament to the informant's charisma. Volz shared the concern but saw denial, too. There was responsibility to go around. The one person Volz didn't blame was the attorney, Micki Brunner. The prosecutor had done what she could. If their snitch sounded like a liar with her, defense attorneys would have chewed him up in court. Tobin hadn't given her much choice.

Brunner in the end still salvaged a conviction, in part thanks to another case worked by Severtson and Volz. Volz had prepared search warrants for Henry Narte, the diver who had been featured in the *Los Angeles Times* story, and was pulling the rest of that case together. Brunner indicted the fisherman and his crew for funneling four hundred thousand dollars' worth of poached clams to DeCourville and another buyer who sent them overseas. Narte had made 191 illegal shipments and would ultimately be given a five-year prison sentence. But the cops' work tracking those sales also gave Brunner the leverage she needed to convict DeCourville. Brunner rattled DeCourville enough that he pleaded guilty to extortion and trafficking. She chalked up a conviction without ever needing Tobin. In early 1998 DeCourville would begin a forty-month sentence in federal prison.

The other cases involving Tobin were lost. Canfield, the bomb maker, and five other bit players would never face federal charges. Severtson pushed Agent Dali Borden to make at least some of the allegations stick, either using what she had without Tobin or by getting more information. But everyone else saw the odds against her. Borden would never get the information she had again without having someone inside.

Detective Kevin Harrington fumed. He'd wasted a lot of time investigating Canfield, and now it would go nowhere. He stormed around, swearing about the Feds. Harrington knew Tobin's history was a problem, but he didn't think it should have been enough to kill the case.

Someone could have made more undercover calls, he said. Something else could have been done.

The rest of the geoduck investigators were still busy. Wildlife trafficking cases had only multiplied while they had been chasing after DeCourville. The agents investigated three separate groups, each of which took contraband valued above six figures. Agents Andy Cohen and Al Samuels and another state detective took on another peculiar and unrelated group of divers. One of these divers feared telephones. One threatened to force-feed dog food to his enemies and bury them in a backyard pit. Another suspect was the son of Officer Obie from the song "Alice's Restaurant," who arrested Arlo Guthrie for littering and kept him out of Vietnam. William Obanhein's son and the team of poachers fished for clams from a boat with a pickup camper hanging three feet off the stern. The fishermen draped carpeting around the back to hide divers slipping into the water. It was the most obvious poaching operation the agents had ever seen. Tracing the sales proved more complex.

After state detectives came across the boat on the water they searched the boat and the crewmen's homes. They found three handwritten invoices that were addressed with fake names and addresses but included a working pager number. At the airport they searched through thousands of bills until they found others listing the same number. Obanhein told the agents the divers sold to a buyer in Brooklyn who bribed customs officials to get the clams into Hong Kong. Agent Cohen spent part of June in New York chasing information, while Samuels worked the case from Seattle. They eventually caught six men stealing twenty-three tons of clams. In a separate case he worked with Detective Volz, Cohen chased another six geoduck thieves to Boston. Those two cases alone had involved nearly $2 million worth of shellfish.

The DeCourville investigation seemed in some way to have sapped Severtson, who increasingly confided in Volz that he was restless. His career had stalled. He did less investigating and more fieldwork. He castigated boaters for harassing killer whales. He responded to calls about sick seals on a beach. He dashed off an angry letter after a lawyer griped that Severtson's cases were too *small*. This was not him. The old Severtson turned over rocks relentlessly and practiced evasive maneuvers in his free time. This Severtson was a bitter shell. During an interview with a suspect during this period, Volz glanced over at Severtson. He found the agent dead asleep.

Doug Tobin, on the other hand, had been hustling to make his mark, both as an entrepreneur and as an artist. The year since he'd begun working with the cops in July 1996 had been a whirlwind. Tobin tended to launch new ideas in bunches and at lightning speed, but they usually fizzled. During the last year, though, even some of his grander dreams took root.

Tobin spent hours carving, which gave him a creative outlet to focus his chaotic perceptions. He displayed wood sculptures at festivals and fairs. Tobin worked in the Salish tradition, characterized by circles and swirling crescents, and the colors red, white, blue, green, and black. He claimed he could look into a slab of wood and see internal forms. He cut masks of frogs and oval-headed wise men, dug flowing feathers into wishbone-shaped driftwood, and shaped and painted paddles and canoes. Tobin showed his work in Olympia galleries and gained prominence at Native art fairs. Wealthy collectors bought his work, as did the Boeing Company and the state ferry system. More than once he declared himself among the best carvers in the world.

In mid-spring 1997, he won the type of lucrative commission he'd long hoped for. A local maritime group wanted the region's rich history represented in an elaborate art piece—something to celebrate nature, European and Native culture, and the meeting of land and sea.

The Port of Olympia, which ran southern Puget Sound's marine cargo terminal, would pay sixty-six thousand dollars to a local artist to carve a Salish welcome pole from ancient cedar. The pole would tower over the water from a new public plaza, just a few miles from the Washington State capitol dome. Tobin eagerly sought the slot and got it.

Tobin bubbled with images and ideas for the commission. He drove out to a studio run by Duane Pasco, a powerhouse of Northwest art whose wood sculptures could be seen everywhere—national parks, Disney's Wilderness Lodge in Florida, a Johnny Depp movie, the Neiman-Marcus Christmas catalog. Years earlier Tobin had convinced Pasco to mentor him. Now he sought Pasco's help again. Tobin's ideas spilled out in a jumble. He envisioned lapping waves and leaping salmon, a moon face, a killer whale, harmony and dichotomy, tradition, and expansion. Pasco didn't bite at first. Tobin had energy but only half-formed concepts. Pasco knew Tobin had a thing about wanting to be the best. He feared Tobin just wanted to associate with a big name.

He told Tobin he was too swamped to work for free, but when Tobin yanked more than two thousand dollars from his pocket, Pasco changed his mind. The two men and a Pasco protégé sketched some patterns. The pole would include a canoe, a schooner, a seagull. Mother Earth would encircle images of man and woman. Tobin would later add a geoduck shell. His career as an artist was finally taking off.

Tobin seemed equally triumphant in business. In 1996 he'd met two Canadians, Jeff Albulet and Julian Ng, who marketed crab, sea cucumber, and manila clams through a seafood company outside Vancouver, British Columbia. Ng spoke Cantonese, the dialect of southern China's food capital, Guangzhou. Albulet, a retired commercial-airline pilot and horse-racing enthusiast, had money and could fly trade routes. The Canadians wanted a steady geoduck supply to sell in Taipei and Hong Kong.

The three men met at a friend's suggestion and formed a partnership. Albulet saw Tobin as a wiseass with perfect comic pitch and a generous spirit. Early in their partnership, he watched Tobin shout "Ya hungry?" to a homeless man outside a Red Robin and then throw his arm around the man's shoulders and take him inside for lunch.

In May 1997, just weeks before DeCourville threatened Casey Bakker, Tobin started another business venture. In between spying on geoduck thieves, Tobin partnered with a friend, Adrian Lugo, who ran one of the region's most successful minority-owned construction companies. Growing up, Lugo had moved from one dusty south Texas town to another as his father worked odd jobs, sold vegetables, or fixed cars to pay rent. He served in Vietnam, worked briefly as a painter and sculptor, and put himself through college. Now Lugo owned a construction company that employed sixty people, brought in $20 million a year, and earned a spot on *Inc.* magazine's list of the country's five hundred fastest-growing businesses. He won civic awards for assisting minority workers. Lugo prided himself on being generous, too. He was an entrepreneur, but he was a Christian first and believed in helping others where he could.

A mutual friend years earlier had introduced the two men when Tobin was in need of work. Tobin was fast, disciplined, and unmatched with a bulldozer, and he impressed his new boss with his eagerness. Tobin followed Lugo around and asked about the nuances of his business and how Lugo had made himself so successful after starting with nothing. Lugo saw initiative worth encouraging. When Tobin tracked down Lugo again that May, he arrived with a proposition: Native American divers made less than whites selling geoducks, but as a licensed seafood dealer, Tobin could pay Indian divers more, and in cash. He could draw most Native divers and build a bigger, steadier geoduck supply with larger profits. They could package clams themselves and sell finished product to the Canadians Albulet and Ng. It would be good for Lugo and good for Indians. Though he grilled Tobin, Lugo was won over. The two talked for hours about making profit and helping out hardworking souls. For Lugo it seemed a Christian way to do business.

They named the corporation White Duck. Tobin and Lugo rented a thousand-square-foot warehouse in Fife. Lugo Construction fronted the cash to pay divers. Because Tobin would be diving and sluicing through the Sound gathering shellfish, Tobin convinced his partner he needed a bigger, better vessel than his arthritic boat *The Judge*. The bank had just repossessed a sparkling twin-engine crab boat packed with a sixty-four-mile-radius radar, GPS plotter, ship-to-shore radio, an air compressor, water pump, and dive package. With Lugo's capital, Tobin took a trip to the coast and returned with that vessel, his new baby: a forty-two-foot fishing boat, *The Typhoon*.

They transformed the warehouse into a fish processing plant. Lugo's workers hung drywall and painted and installed refrigeration units, stainless steel sinks, and tables. Julian Ng visited the facility from Canada. Health inspectors showed up to ensure it met health codes. Halfway through an already great year, Tobin was poised to make a killing.

It went bad almost immediately. Lugo noticed within weeks that business wasn't exactly booming. He sent Tobin out with thousands of dollars, expecting him to come back with big returns. Instead Tobin would come back with less than Lugo had given him. Tobin blamed his staff and a side partner, a glad-handing former tractor salesman, who Tobin said pocketed the company's profits. Tobin even called in the wildlife cops. Unfortunately, so did the tractor salesman. He told them Tobin had ripped *him* off. To the cops the two men sounded like squabbling children. The men agreed to part ways, so the cops left it alone.

Lugo listened to Tobin's excuses for weeks before making his decision to get out. "I felt like I was in a scene from *The Producers*," Lugo later said. If someone stole from Tobin, that was Tobin's problem; Lugo wanted the money he'd invested. He filed liens against *The Typhoon* and threatened to padlock the warehouse. When the Canadians finally called, offering to buy Lugo out, pay off the liens, and take ownership of *The Typhoon*, Lugo was relieved. After a tough round of negotiations with the Canadians, Tobin even urged his new partners to toss Lugo a little extra.

"Don't worry, pay him," Tobin said. "You know the kind of money we're going to make."

Ed Volz could feel the Feds' priorities shifting. Severtson and his agents were wading into new issues, and Volz feared that would leave him and his fellow detectives alone dealing with shellfish poaching. Severtson was moving on to a more high-profile assignment, helping police the nation's first whale hunt in more than half a century.

For thousands of years the Makah Tribe on Cape Flattery, at the far northwest tip of the Olympic Peninsula, had hunted barnacle-encrusted humpbacks and Pacific gray whales. Gray whales rebounded when commercial hunting stopped in the 1920s, and by the 1990s they were no longer an endangered species. The Makah sought a special permit to resume whaling from canoes. Word of a pending hunt spread as far as Israel and Japan. Antiwhaling groups and animal-rights activists flocked to America's wettest corner to try and scuttle the hunt. By piloting jet skis, boats, and planes, they would try to get between eight tribal hunters and a great gray whale. Tribal police were there to handcuff protesters and confiscate their rafts while several hundred armed state and federal cops waited on the cape, fearing an explosion between the tribe and protesters.

Severtson served as a mediator, but his sympathies lay with the antiwhalers, particularly with Captain Paul Watson and the Sea Shepherd Conservation Society. Watson, large and white-haired, sailed the seas like an eco-Rambo. He hounded sea cucumber poachers in the Galapagos, rammed Japanese whalers, and tried to convince a Las Vegas casino tycoon to buy a submarine that would spook sea-turtle poachers. One of the founders of Greenpeace, he was voted out in 1977 because he was seen as too aggressive. The previous year, a Norweigan court had convicted him in absentia of trying to sink a fishing boat after Norway defied an international whaling ban. Captain Watson policed the seas in ways Severtson wished he could, unburdened by bureaucracy. Watson arrived in Neah Bay in a navy peacoat and jeans and carrying a bullhorn. Severtson brokered meetings between Watson's Sea Shepherd Conservation Society and law-enforcement groups. He tried to help show other cops that Watson's group didn't intend to spill blood.

Severtson's enthusiasm was back. Watson called him "the first federal agent who really worked with us as a straight shooter," but that didn't improve Severtson's relations with his bosses. Watson had been jailed in more countries than most of Severtson's colleagues had visited. Many federal cops considered Watson an eco-terrorist. Severtson called him a friend.

Volz juggled complicated emotions watching it unfold. Government work, especially police work, was inherently political, and Severtson had spent thirty years refusing to get along. Good for Rich for sticking to what he believed, Volz thought, but why intentionally antagonize the brass? Severtson's obstinance kept hurting him. He was one of the best cops Volz knew, but the agency would never put him in charge in Seattle, the only job he had ever really wanted. Those thoughts were jumbled in Volz's head in the winter of 1998, when he happened into Severtson's office and saw an announcement on his desk. Volz picked it up and Severtson reddened. The federal agents—Andy Cohen, Al

Samuels, Dali Borden, and Severtson—would be called back to Washington, D.C., and feted by their superiors. They had been awarded the highest honor bestowed by the U.S. Department of Commerce: gold medals for service. The secretary of commerce granted the awards for "rare and distinguished contributions of major significance to the department, the nation, or the world." In Severtson's case it was a truly rare tribute—he was one of the few who could now claim to have received the award twice, the first time for fighting salmon poachers in the 1980s. The new medals applauded the agents for their multiyear undercover investigation into the illegal harvest and sale of more than a million dollars in geoduck clams. The citation pointed out that "they discovered that this shellfish was illegally marketed through back channels of interstate and foreign commerce to customers in Canada, Japan, and Hong Kong. They uncovered organized criminal activity which threatened the health, welfare and safety of the public." The agents "removed tons of contaminated seafood products destined for American and international markets. During the investigation, they exposed and terminated the largest illegal bomb factory in Washington State history."

Volz cringed. The Feds were getting an award? Certainly the cases that agents Cohen and Samuels were finishing were complicated and important and appeared to be going smoothly. But many of those cases also involved state agents, and then there were the others that Tobin had helped botch. Volz knew how close they'd come to losing DeCourville. He had no expectation that as a state employee he would win a federal government award, but Severtson had clearly gone out of his way to keep his award quiet.

By 1999, the Feds had moved on. Assistant U.S. Attorney Micki Brunner was pulled into a long investigation of pipeline companies after a fatal gas-line leak near the Canadian border. Two ten-year-old boys had been playing with fireworks in a park near a creek on a summer day when one struck a butane lighter and sparked an explosion that sent a black cloud soaring six miles high. A pipe had ruptured and spilled a quarter of a million gallons of gasoline into the creek. Both boys died in the hospital the next day. The blast also killed an eighteen-year-old fisherman. Brunner was assigned to find out if the accident was the result of criminal negligence.

Severtson retired and began teaching law enforcement at a community college north of Seattle. He spent his leisure time bow hunting in the mountains. Dali Borden felt passed over, too, after being denied a shot at leading operations in Seattle. She took a job as a cop with the Food and Drug Administration, investigating food-safety cases from mad cow disease to pill tampering. Andy Cohen moved to Boston and rose within the agency. When Severtson left, Al Samuels took the helm and changed direction. Four federal agents had spent several years looking into the theft of giant clams. Samuels knew that they hadn't rooted out corruption in the geoduck industry, but the agency needed to focus on other cases. There were a hundred other commodities to investigate. Federal resources were not endless, and there were many other fish being smuggled from the sea.

The geoduck world was changing, too. Since the 1970s, biologists had been searching for ways to farm geoducks like crops, which could help protect wild geoducks and bring consistency to products and markets. Scientists tried getting clams to reproduce in buckets and in hatcheries and in nature, planting them on beaches and dumping them haphazardly from boats, but the crop always ended up dying. By the mid-1990s, scientists had cracked a few basic riddles. Shellfish companies began rearing geoduck larvae as small as the half moon on a man's pinky nail in tanks as big as brewery boilers. Clam farming was on its way, but it would be years before it produced geoducks in real volume, and when it did, it would stir controversy of its own. In the meantime seafood lovers still wanted wild geoducks and wanted them enough that people were quick to steal them.

Twenty-one people had already faced federal felonies, two had fled the country as fugitives, and several million dollars' worth of shellfish had been stolen. But since the Feds had made clear they would no longer police geoducks, all the responsibility for clam crimes fell to state detectives. Volz's superiors decided they needed a more systematic approach. Rather than just investigate individual crimes, Volz's chief urged a handful of undercover officers to spend time documenting general trends within the geoduck industry. They would take a more holistic approach to understanding what was happening within the Northwest's high-dollar fishery.

The detectives watched geoduck harvests and became better acquainted with the fishermen. They visited the docks and ran into regulars and spoke with everyone they could think of who might offer insight, including Doug Tobin, who had left word with the detectives that he had information. He wanted to meet to talk about it in person. That fit perfectly with the detectives' own plans.

On a cold morning early in 2000, Volz called Tobin into his office. The detectives wanted Tobin to describe on videotape how things worked in the field, how poachers ran roughshod over those who followed the rules.

Tobin showed up looking haggard, his face swollen with exhaustion, a paper cup of coffee in his hand. He wore a purple-and-black down jacket, his flannel shirt open to midtorso. That was just fine; he didn't need to look pretty. The detectives only needed him to start talking. Tobin spoke for hours for the camera, outlining what he knew about supplying geoduck to the world.

Tobin brought a yellow legal pad, complete with a suggested list of suspects to pursue. He told Volz about two men who overharvested four hundred pounds of clams a day and about a buyer he suspected did everything under the table. He mentioned a Canadian company shipping stolen geoducks by truck to California, and another that used shellfish as a front to smuggle untaxed cigarettes mixed with seafood. He said the man who had invited friends over to watch surveillance footage of himself stealing clams was back on the water, breaking the law again. And another longtime player was shipping clams by truck to markets in San Francisco and Los Angeles, about two thousand pounds each week. Tobin once more offered to sell geoducks undercover to help catch these crooks.

Volz stayed silent. He had other ideas.

A few weeks later, Volz dimmed the lights over a scrum of wildlife officials, including his chief, a former head of the Washington State Patrol, and the leaders of two natural resource agencies. Volz and the other detectives had prepared for weeks for this moment. Since the Feds had decided geoduck fishing was a state problem, they hoped to get more attention from legislators and rule makers by showing them how out of control clam fishing had become. Volz pulled the shades, then turned on a projector and his laptop. A headline flashed up on a wall: "A Resource in Peril."

Volz flipped through slide after slide, recounting the half-dozen smuggling operations he and other cops had busted up. Special Agent Al Samuels's investigation of the Brooklyn seafood dealer. The case Volz and Agent Cohen had followed to Boston. DeCourville. The room stirred, but Volz pressed on. He described the way his special investigations unit worked, then flipped to a slide he hoped would be the day's showstopper: a videotaped interview with a geoduck insider.

The speaker's face was blurred and his voice was altered to protect his identity, but his presence was commanding: flannel shirt open to an ample belly, revealing a whale-tooth necklace, and arms stuffed in a down jacket. "I've been a diver, and a buyer, and a packer, and I basically ship geoduck all over the world," Doug Tobin said. "When I got into this industry, I was as green as a pool table and twice as square. But I've seen corruption from day one."

Volz watched alongside the others like a coach sharing highlight reels. The day Volz had filmed him, Tobin had come to the office late, tired, and unprepared, his eyes puffy. Volz had edited out scenes of Tobin offering details about open cases. But now he watched as Tobin detailed on the fly the many ways poachers hid their work: Boat operators stashed clams in skiffs that sped off before the catch was tallied. Brokers forged paperwork to sneak loads in and out of Canada. Lately divers were attaching water hoses through their boat hulls to obscure evidence of someone prying clams down below with a jet sprayer. Divers increasingly used scuba rebreathers, which did not release bubbles that could be seen from the surface. "There's no noise," Tobin said, shrugging to indicate his dismay. "They could be anchored right off Alki," West Seattle's crowded recreational beachfront. "If this building was sitting on a beach somewhere, close to the water, they could harvest 'ducks right in front of your office and you wouldn't know."

There was a lilt of sadness in his voice. "I feel right now there is an incredible virus, kind of running rampant," he said. Tobin offered his own family's story as an example of how mismanagement had squandered the Northwest's natural bounty. He spoke of his father's business, Alpine Construction, and how well his family had once done cutting down trees and fishing. In the days of plenty, people took their share and nothing more, and everyone made money, Tobin said. There was something left for the next generation. Now the biggest old-growth trees were visible only in the faded black-and-white photographs hanging in area restaurants. Without strong action to protect the geoduck, the region would be left with nothing but its stories.

"I've seen the demise of old-growth timber, to where our standards are today. I was also unfortunate—or fortunate—enough to see the demise of the salmon. With the logging, at one time, you could raise your family, and my dad did incredibly well. And we did incredibly well on salmon. It used to be that you could raise a family just on salmon. So, in my book, we've basically lost the heart of both of those. And I see the same demise being conducted with geoduck." Not everyone, Tobin claimed, saw clams as wildlife worth protecting. They saw only the millions upon millions of dollars in the sand. "If something isn't done, it's going to be like the timber and the salmon," he said. "The party's over. We either do something, or we lose."

chapter nine

A SEA OF ABUNDANCE

Doug Tobin hadn't let inconvenient facts cloud his story—to suggest geoducks were in jeopardy was ridiculous, and everyone knew it. Puget Sound holds millions of geoducks. More than 165 million inhabit waters at fishable depths, according to official estimates. Millions more are presumed to live past the seventy-foot limit for commercial clam diving. Researchers using deep-sea cameras have captured fields of geoducks 350 feet below the surface.

And yet Tobin couldn't entirely be dismissed. Geoducks are unusual creatures that reside in a complex environment. They live so long and grow so slowly, they are less like fish than trees, an irony not lost on shellfish biologists. Most people presume abundance alone is enough to ensure these ancient clams will thrive. But the same had been said of the Pacific Northwest's salmon and old-growth forests. By the time Ed Volz made his presentation, Puget Sound chinook had dwindled to such low numbers they needed the protection of the Endangered Species Act. And lumberjacks had mostly been banned from the Northwest woods. After a century of logging, scientists had discovered that hacking down so many thickets of Douglas fir and cedar threatened the entire forest system. Timber harvests put at risk other life in the woods: northern spotted owls, thumbnail-size beetles, silver-haired bats, spatula-shaped mushrooms, and scurrying red tree voles. The Pacific Northwest had gone to great lengths to save those species, but by the winter of 2000 many were still in trouble. The forests that remained were just fragments of what they'd been.

Geoducks are the old-growth forests of the sea, but digging for them had been managed more conservatively from the start—fishermen were never allowed to take anywhere near as many clams as loggers took trees. They were restricted by law because scientists understood that the creatures' extraordinary lifespans made the consequences of removing them tough to grasp. Thirty to fifty million in any given year were off-limits solely because they'd been contaminated by pollution or red tides. But rules protecting creatures from overharvest do little good when they're not followed. The margin for error is slim for long-living marine creatures, and Puget Sound faced new ecological challenges. In a world beset by pollution, climate change, and acidifying oceans, massive poaching operations just complicated everything, and the marine world already was complicated enough. Scientists trying to gauge the long-term sustainability of geoduck fishing found answers elusive because they rested on a more fundamental query: How well did we really understand the sea?

History is filled with once-prolific sea creatures humans thought would last forever. The first of those appears to have been a clam. Long before the earliest people worked their way to the Red Sea, *Tridacna costata,* a huge, squiggly jawed shallow-water clam, was among the Middle East's most plentiful marine animals, accounting for 80 percent of that region's giant shellfish. Then humans showed up 125,000 years ago and discovered this accessible seafood treat. Soon after, the number of giant clams plummeted. *Tridacna costata*'s decline is the earliest known example of the overexploitation of marine life by humans. It wouldn't be the last.

Stocks of the world's most prized fish, Atlantic cod, collapsed in the 1990s in the world's richest fishing grounds, bringing the risks of overfishing to the world's attention. Fishermen for centuries had swept nets back and forth across the submerged plateaus of the North Atlantic east of Newfoundland and east of Cape Cod, places with names like the Grand Banks and Georges Bank. They scooped a bounty of cod for fish and chips or fillets. Then technology, advances in fishing trawlers, and the rising popularity of consumer staples like frozen fish sticks sent the catch into the stratosphere, and cod populations crashed. A single ship by the 1980s could net more cod in a day than an entire fleet centuries earlier could land in a season. By 1994, cod stocks off Georges Bank had plummeted 40 percent in just a few years. The government declared that fewer captains would be allowed to fish, and they would get to spend fewer hours at sea. Two years later, cod stocks dropped more.

Still, invertebrates such as shellfish, sea urchins, and sea cucumbers had once been thought naturally resistant to overfishing. In part that was because they were fished by divers; some number of the animals would always survive in deep water far beyond where divers fished. If fishermen took too many, regulators shut down seasons for a few years, believing wild creatures would rebound on their own. But Puget Sound's pinto abalone already were demonstrating that nature rarely follows human rules.

Abalone had been popular as a delicacy long before poacher-turned-informant Dave Ferguson started stealing them. But the rise of Asian markets in the 1980s and 1990s helped generate a new boom in demand. Connoisseurs pounded the meat with mallets, boiled it in soup with cabbage, stir-fried it with pepper and garlic, or served it as steak stuffed with prosciutto and pine nuts. By the end of the century, illegal fishing threatened abalone around the world. Poaching by Chinese gangs in South Africa was sparking gun battles in Johannesburg. British Columbia authorities would soon employ shellfish-sniffing dogs on ferries and use a DNA database to track abalone poached from their waters. One team of crooks would be caught trucking eleven thousand abalone off Canada's coast. Smugglers regularly took hundreds of thousands of dollars' worth each year from California, which holds the world's last remaining populations of red abalone.

Washington had never allowed commercial fishing for pinto abalone. But by the time of Ferguson's arrest in 1994, the creatures already had begun disappearing. Biologists knew poachers almost certainly were to blame, but they were eager to better understand the problem. For years, they dived along the San Juan Islands and investigated, and through the 1990s, they noticed the decline actually picked up speed. But the more researchers learned, the more confused they grew.

Abalone are voracious grazers, and unlike geoducks, they move about underwater. They cruise at night, scraping algae off rocks and feeding on drifting kelp. That ability to wander is how they survive. They must move extremely close to one another in order to reproduce. But that clustering makes them easy prey for thieves, who can scoop them up by the dozens during a single scuba dive. Yet while scientists found that abalone in Puget Sound were growing rarer, the ones that remained were also getting bigger. If poachers were the problem, that didn't make sense. Wouldn't poachers want the biggest ones?

Then the researchers hit upon a troubling answer: Average abalone sizes were getting larger because a greater percentage of the shellfish were older. That could mean only one thing. The shellfish were living longer but not being replaced by offspring. The abalone thieves in the

1980s and early 1990s had created an even bigger problem than first thought. So many of the shellfish had been removed from Puget Sound that the remaining abalone lived too far apart to congregate. The creatures that survived weren't getting close enough to mate. Poachers years earlier had sent Puget Sound abalone into a death spiral. Not even the biggest ones were plentiful enough to make serious poaching worthwhile anymore.

About the time Ed Volz was making his video presentation of Doug Tobin, scientists were struggling to understand poaching's effect on geoducks. They started by studying geoduck mating. Several times from January to June, male geoducks exhale sperm in billowy puffs from their siphons, prompting females to release millions of eggs into the water column. Researchers call this broadcast spawning. Corals do it, as do some oysters, spreading their seed in milky bursts, some of which can be seen from satellites. Egg and sperm meet in the water, and within days shelled larvae begin to swim about. Nearly all of these infant geoducks are consumed by other marine life.

The geoduck compensates for this poor rate of survival with a robust libido that allows it to procreate for a century. The baby clams that survive this open-water gauntlet tumble to the seafloor after just a few weeks, remaining exposed until they dig safely into their burrows, where they can finally live unscathed by predators. Sea otters and starfish occasionally tunnel into these hideaways, and crab or spiny dogfish might graze on exposed siphon tips. But while other clams must scoot away to avoid hungry fish, a geoduck merely retracts its neck like a turtle. Most geoducks get to live in peace, grow old, and expire.

Yet geoduck populations fluctuate wildly, and no one really understands why. Once, in an area of Puget Sound that had never been fished, a scientist randomly pulled out a thousand wild clams. Thirty percent of them were exactly the same age. For an animal that can live 150 years and spreads its seed randomly many times each year, the odds against that happening should have been astronomical. But the giant clams appear to reproduce in spasms. Some years the creatures are monstrously fertile, producing an exceptionally large crop of baby mollusks. In other years offspring survival is horrific. Scientists have no clue as to what most influences reproduction and continued existence. It could be food or water temperature or tides or wind. Regardless, the rate at which baby clams survive to adulthood dropped in half after the Great Depression, even though commercial geoduck fishing didn't start until 1970. Then it picked up again.

That's part of the reason divers each year can legally take less than 3 percent of the clams at fishable depths—a rate scientists presume is sustainable. To know for certain, they need an accurate count of geoduck populations. So, since the late 1960s, researchers in scuba gear have investigated the Sound each year. Working two by two in straight lines, they move row by row and bed by bed, slowly counting the geoducks they see. Once the clams are tallied, biologists log their totals in a database they call the Geoduck Atlas, which they subject to mathematical modeling. That helps them recommend where and how much to let divers fish.

But this strategy is only as good as the raw data, and there are significant gaps. Clam counting is slow and arduous, and once a bed is mapped scientists rarely waste time going back. It might be twenty years before an area is revisited. And when clam beds are closed to fishing, regulators assume the geoduck population stays the same. Often that's not the case. Thirty years of massive poaching, combined with basic gaps in knowledge, can undermine their assumptions. "If something happened to those clams—if they died or someone fished that bed illegally—we wouldn't know it," said Washington Department of Fish and Wildlife shellfish biologist Alex Bradbury.

Many things can happen to those clams. Climate changes and environmental damage already were unraveling some of Puget Sound's natural systems in ways scientists were still just coming to understand. Logging, warming seas, and pollution from leaky septic tanks starved creatures in Hood Canal of necessary oxygen. Anchovies, sand lance, cucumbers, octopuses, perch, and other fish sometimes suffocated in mass die-offs. When scientists finally surveyed a section of Hood Canal's geoducks, they found that roughly half were dead.

There was also the looming threat from ocean acidification. Seawater typically is slightly alkaline, but when oceans absorb carbon dioxide from the atmosphere—as they have by hundreds of billions of tons since the Industrial Revolution—the waters become slightly more corrosive. Climate modelers in the 1990s predicted greenhouse gases would lower the pH of marine waters by the year 2100. They expected to see it first in parts of the Pacific Northwest, where waters already are more acidic. But scientists would find evidence that ocean acidification appeared to be accelerating and altering the chemistry of the seas far sooner than expected. The animals most susceptible to these minute changes: shellfish. Baby clams and oysters are particularly at risk of erosion from less alkaline seas, which can completely dissolve their protective calcium carbonate shells.

Biologists also worried about waste from another type of illegal fishing, a practice called high-grading. The difference in market price between smooth gleaming-white top-quality clams and stubby dirty specimens could be several dollars a pound. The only way divers could tell which they had was to dig them up and take a look. Many pulled out clams, took a look, then illegally tossed the bivalves aside. Removed from their beds, all geoducks die. The question for researchers was: How often did it happen?

In 1999, biologists had accompanied the detectives on an unusual undercover operation. They hung out onshore near geoduck beds, hiding or trying to look inconspicuous while commercial divers worked nearby. When these divers left at sundown Volz ferried the biologists in an undercover boat out to where the fishermen had been working. The researchers slipped into their own scuba gear and counted the dying clams they saw loose on the seafloor. They also tallied the empty shells after the season ended. What they found was startling—divers in just one location had discarded seventy thousand pounds of unwanted clams. On some beds, scientists concluded, illegal waste probably added 30 percent to what regulators presumed was harvested each year.

Predicting how poaching might change the population of marine creatures is almost impossible. Illicit activities are by nature hard to quantify. Just the number of geoducks people could prove had been stolen would add another 10 percent to the annual catch. And as the DeCourville case showed, cops usually caught poachers stealing only a fraction of what they actually took. That means the calculus used to set sustainable fishing rates is based on numbers that may be mildly— or wildly—inaccurate. Scientists understood this and tried to adjust accordingly, but that involved a fair amount of guesswork. And there was no accounting for the shifting ecology of the seas. Geoducks certainly would remain abundant for some time, but no one could predict how long that would be the case.

By the turn of the century, overfishing and seafood smuggling were commonplace. The seas' biggest fish—sharks, tuna, marlin, sailfish, swordfish—were disappearing, and scientists around the world struggled to trace illegal fishing's role. Perhaps $1 billion in fish caught illegally from the high seas would soon be sold each year in European markets. Fishing pressure was rising faster than the ability of modern science to comprehend the marine world's complexity. Poachers and smugglers increasingly made off with fish species that researchers had never studied much at all, and not just fish species chefs served for dinner. Fish of all descriptions—big, small, tropical, ugly, razor-toothed, even poisonous—were being taken illegally, often for purposes that had nothing to do with food. And just as researchers were learning with geoducks, it was impossible to really understand the consequences, even among those creatures that were long presumed to be abundant. Nowhere was that phenomenon playing out more strangely than along the California coast.

The trouble came to light on May 15, 1991, during Kenneth W. Howard's fourth dive of the day. It would be the last of his life. At seventy-three feet, something went wrong. Howard untethered himself from his air line and began sucking an emergency oxygen bottle. It ran out before he reached the surface. Authorities never learned precisely what happened, but the accident would reverberate through the public consciousness. It wasn't the fact of Howard's death that the public seized on, but what Howard had been doing when he died.

Howard was one of dozens of people who regularly dived the east side of Catalina Island off the coast of Southern California to collect fish. The thirty-four-year-old knew well the fertile stretch near Isthmus Reef where divers moved among moray eels, lobsters, and the blue sprigs of Christmas tree worms. It was a popular inlet, a fish-rich scuba-diving paradise, but Howard did not use scuba gear. A partner fed Howard oxygen through a hose from a compressor on board a boat while down below Howard and other divers bagged tiny snails and baby octopuses, stingrays, eels, orange Garibaldis, pipefish, and sharks. The divers sold their catches to brokers in the marine aquarium-fish trade, who in turn sold them to pet stores around the world. Howard typically sold fish for $5 to $7 apiece and often pocketed $200 a day.

On an exceptional day the week before his death, he caught forty-eight baby leopard sharks—creatures popular enough to command a premium. Howard walked away with $654 for a day's work.

Howard's job as an aquarium-fish wrangler caught California citizens and the government by surprise. Few had heard of aquarium-fish smuggling. Freshwater aquariums around the world were stocked with fish raised mostly in captivity, but the $300 million saltwater marine-aquarium trade was almost exclusively supplied by wild fish. Most came from waters off Indonesia or the Philippines, but fishermen worked this trade all over the world, including Florida and Hawaii. In California in 1991, that type of fishing was legal and entirely unregulated. The California Department of Fish and Game had no idea how many aquarium divers there were, or how many fish they took. Authorities who investigated Howard's death learned that other divers had been working the bays off Catalina since the 1970s, using makeshift vacuum hoses and plastic tubes attached to motorized generators ("slurp guns"), which they jammed into reefs and rocky crevices. These devices sucked up all nearby fish, even though the aquarium business targeted only a few species.

Howard's accident alarmed those who cared about marine fish. "The death of that diver opened Pandora's box," the head of a Catalina Island conservation group told the *Los Angeles Times* in 1992. "We had no idea that kind of commercial fishing was going on or how destructive it is. They're raping the ocean, taking everything they can catch."

Biologists turned to state lawmakers for help. That year the California legislature banned fish collecting around the reefs of Catalina Island. Up and down the California coast, aquarium-fish gatherers now needed special licenses and were required to document their catches. Lawmakers named the Garibaldi California's state fish, and collecting it, too, was soon prohibited. The state also outlawed catching or possessing baby leopard sharks, which were extraordinarily popular with aquarium owners. Thin and bendy with gray-and-white skin, leopard sharks are graceful and docile and roam coastal waters from Mexico to Oregon. They can grow up to seven feet but age slowly. Aquarium owners bought young pups between eight and forty inches, and when the fish outgrew their tanks, homeowners tossed them down the drain. After Howard's death, researchers realized that practice was dangerous to the species. "Removing too many pups could have huge significance," said Suzanne Kohin, a shark expert with the National Marine Fisheries Service. "We have such poor data on these types of sharks that if someone removed a bunch of them, we might not notice the decline for years." In 1994, California lawmakers outlawed catching or keeping leopard sharks smaller than thirty-six inches. The law let fishermen catch adult sharks but protected the babies needed to sustain the species. At least that's what lawmakers thought.

Several years later National Marine Fisheries Service special agent Roy Torres heard from a colleague in Miami that Florida wildlife detectives had raided a pet dealership that had sold eighteen baby leopard sharks. The pet distributor got the sharks from a supplier in Los Angeles not far from Torres's office in Pacific Grove. The investigation also turned up the name of a second supplier: Kevin Thompson of San Leandro, a small suburb on San Francisco Bay. Thompson had shipped boxes marked LIVE TROPICAL FISH to Florida a year earlier. Thompson was apparently a regular shipper, and Torres spent the summer learning about him.

Meanwhile, that spring, the U.S. Fish and Wildlife Service in Illinois received a tip from an angry pet dealer that a rival near Chicago was illegally selling baby leopard sharks. When a federal agent visited the Chicago shop the owner and his wife confessed they'd just received 101 leopard sharks from California sent by Kevin Thompson. Torres staked out Thompson's address and noted an odd decal in a window of the house, a line-drawn symbol he had never seen before: a yellow cartoon depiction of a family. He saw the same sticker in a window down the street, at a brick building known as the Bay Area Family Church. It was the symbol for the San Francisco–area chapter of the Reverend Sun Myung Moon's Unification Church. Kevin Thompson was its pastor.

The Reverend Sun Myung Moon, leader of the Holy Spirit Association for the Unification of World Christianity, may be most recognized for presiding over mass weddings, his ownership of the *Washington Times,* or his conviction on tax-fraud charges in the early 1980s. Less visible is his ownership of vertically integrated fishing and seafood businesses.

The Korean-born Moon considers himself a messiah, and in the mid-1970s he urged his disciples—"the Moonies"—to dominate the seas. The Moonies bought land in many of America's busiest fishing ports: Miami, the Gulf of Mexico, Norfolk, Kodiak, San Francisco Bay, and the cod-fishing village of Gloucester, Massachusetts. They bought shipyards, fleets of tuna-fishing boats, and seafood processors and launched what would become the country's dominant sushi distributor, True World Foods. In one sermon Moon christened himself "King of the Ocean" and outlined his vision. Land would soon be inadequate to feed the world, but the ocean was boundless and everything in it was edible. "I have the entire system worked out," Moon said. "After we build the boats, we catch the fish and process them for market, and then have a distribution network. This is not just on the drawing board; I have already done it. All these ideas I conceived many years ago, and I knew just where to start—with tuna fishing—because it is the essential way to train people in the fishing spirit."

Kevin Thompson had worked his way to San Leandro in the 1980s from the tuna industry and a Moonie church in Gloucester. Thompson was a doughy, amiable Brit raised in Manchester, England, and spoke with a disarming Scottish brogue. Towing his twenty-eight-foot boat called *One Hope,* he moved west to serve as the Bay Area Family Church's youth leader and set about instilling Moonie values in children through fishing.

The parishioners called their sanctuary Ocean Church. Years later, after becoming pastor in San Leandro, Thompson would tell his congregation how he stumbled into selling sharks. "Some of you know that Ocean Church . . . has had this little shark business," he said, in a sermon that was recorded and later shared with authorities. "The way we got into that business was totally by accident. Anybody who fishes in the bay knows these little leopard sharks are a pain in the neck. They steal your bait all the time. Fishermen hate them. So for years, we'd take people fishing, and these little sharks would be the pests. So we'd throw them back.

"One day we found out that in pet stores they were selling these little sharks for seventy-five dollars for one shark. These things that we throw away by the hundreds! This one brother who was with us said to this pet store, 'How about if I catch you a few of these sharks? Would you buy them from me?' They said yes, and he went out the next day and he caught five. He took the five sharks to the pet store, sold them for twenty dollars each, and said, 'Here's my number, give me a call. When you sell out I'll get you some more.' That night he gets a call from the pet store. They said, 'We talked to all the other pet stores that we're associated with and we want you to catch us some sharks. And we'll cut out the middleman. And we'll get them for cheaper from you. And we'll make more profit and you'll get some money.' And that's how we started this business, catching little baby sharks."

Thompson's sermon wasn't far from the truth, though he diminished his own role and ignored an important fact: What Thompson had done was illegal after 1994, but he did it anyway. His leopard shark business was at the center of the largest shark-smuggling ring in the country's history. Torres would learn that over many years as many as sixty thousand baby leopard sharks were illegally plucked from California waters and sold into the aquarium trade. Tens of thousands of those sharks, worth more than $1.2 million, came from Reverend Kevin Thompson. For years, his crew stashed fishing gear and stowed stolen sharks in special bins at True World Foods. The sharks were incorrectly labeled as TROPICAL FISH, COMMON SHARKS, or HARLEQUIN TUSK and wound up in pet stores from Texas to Maine. The preacher's outfit took enough sharks to supply international pet dealers in Hong Kong, Germany, Italy, France, Spain, the Netherlands, and other countries, too.

Even to an eighteen-year-old, it seemed obvious that this was wrong.

Brandon Olivia worked and lived with Thompson, who like many Moonie family patriarchs opened his home to lost young men. In 1991 when he came to stay with Thompson, Olivia roomed with another teenager, John Newberry, the nineteen-year-old son of a British policeman. The two young men got on nicely. They took church members fishing and sold flowers on the street. They had no bank accounts and made no money, but they lived for free and bought clothes at Goodwill with money provided by Thompson and his wife. Newberry and Olivia treated Thompson like a father. "They basically did whatever I asked them," the pastor said.

It was with Olivia and Newberry that Thompson learned they could sell the baby sharks they hooked by accident. At the time, doing so was still perfectly legal. Thompson met with a pet distributor in Los Angeles who needed a regular supplier of leopard sharks. Thompson had boats, equipment, and experienced fishermen. Yet Olivia and Newberry struggled at first. They fished under the San Mateo Bridge, using squid or worms for bait, but the sharks they caught were too big and died before reaching Los Angeles. The buyer eventually told Thompson that the sharks needed to be starved temporarily in a large holding tank. That way, the fish could eliminate their recent meals and not poison themselves during shipping.

Leopard shark, *Triakis semifasciata*

Newberry built a handful of twenty-five-hundred-gallon tanks, which he stored behind a True World Foods distribution center. For a while, things went smoothly. Then one day Olivia did not come home from being out on the water. By 8 P.M. the pastor was worried and ran down to the docks, certain something terrible had happened.

Years later, Thompson would recall for his congregation what happened when Olivia finally showed up.

He yelled and huffed at Olivia, but Olivia just said, "I, uh, I ran out of gas."

"You ran out of gas? The law of fishing is you don't go out without enough gas. There's no gas station you can just pull into, right?"

The congregation chuckled as Thompson told his story. Olivia had run out of gas and got stuck in mud at low tide, and he'd kept himself busy fishing while he waited for the tide to rise enough to paddle home. "So I'm mad," Thompson continued during his sermon, "and, you know, I'm giving him my best speech. And then he said, 'But guess what.' And he opens up the lid of the buckets that hold the sharks and he had one hundred sharks in there. The most we had ever caught before that was twelve." Olivia had found the leopard shark pupping grounds, where pregnant females give birth in spring and summer. Soon Thompson and his fishermen had a system. They fished with trout hooks, four to a line, and could catch fifty babies after dawn and be back on land by 9 A.M.

But by then the diver had died off Catalina Island and scientists and California lawmakers were growing wary of the aquarium trade. For every few thousand baby sharks taken, the biologists figured out that the rate at which the species replaced itself actually dipped about one percent, which over a few years could prove quite troublesome. Thompson's buyer in Los Angeles heard lawmakers were about to end fishing for baby leopard sharks. He told Thompson the church should lobby lawmakers in Sacramento. Thompson ignored him, but the buyer faxed the pastor newspaper articles about the pending ban. Thompson ignored that, too.

For years and years after the legislature banned fishing for baby sharks, the church operation continued. When buyers asked if what they were doing was illegal, Newberry would tell them, "Nobody's bothering me."

Only Olivia seemed concerned. He told Thompson after the law took effect that they should do something else. Olivia offered to search for a new way to make money, but Thompson said no. "I was very upset about it but I just lost all power to try and say anything," Olivia later told Torres through tears. "I just couldn't see eye to eye with him about this whole mess. Kevin still was the leader of Ocean Church. And I believe in the underlying fundamental premise of Ocean Church."

Thompson later admitted he didn't leave the business because demand was so high that he couldn't resist the money. On the retail end, his sharks fetched up to two hundred dollars each. He told his congregation that he stayed with fishing in order to stay in the good graces of Reverend Moon, who had been excited to hear about the project. "When I had a chance to tell our founder, Reverend Moon, about it . . . he told me, 'You need twenty boats out there fishing!' He had this big plan drawn out. I said, 'No, no, no, we can't do that.' But he doesn't like the idea that you can't do anything."

No clear evidence ever surfaced that Thompson shared what he was doing with Moon, but authorities were still able to extract a financial penalty from Moon's church, which contributed a half-million dollars to marine habitat restoration in San Francisco Bay. Thompson spent a year in federal prison, and Newberry was sentenced to six months. Four others were also convicted.

Because scientists still don't fully understand the ocean, the precise impact of the poaching may never fully be known. Particularly disturbing is the fact that fishermen targeted the young. "Killing babies before they've had a chance to reproduce is never a good idea," said Greg Cailliet, program director for the Pacific Shark Research Center at California's Moss Landing Marine Laboratories.

But leopard sharks aren't great whites or hammerheads. They haven't been exhaustively researched like other shark species. Scientists who intuitively understand their role in the marine world could only say what shellfish biologists said about geoduck rustling: It would have been far better if the poaching had not happened. "Sharks are the apex predator of the oceans," said University of Washington shark expert Vincent Gallucci. "They're at the top of the food chain. And if you disturb the top of the food chain the ramifications reverberate all the way down." Christopher Lowe, the head of the shark laboratory at California State University at Long Beach, was far more blunt. "Those guys took a lot of shark pups out of the environment. And taking all of those young out *has* to have an effect. We just don't know what it will be."

chapter ten

CRAB MEN

Geoduck enforcement improved marginally after Ed Volz's video presentation. A newspaper reporter on one of the peninsulas across the Sound wrote about the seventy thousand pounds of wasted geoducks biologists had found discarded in mud the previous fall. A handful of editorial writers across the state demanded stricter rules and better policing of clam divers. Over several weeks, wildlife managers monitored geoduck harvests with global positioning systems, underwater video cameras, and unannounced spot checks.

They could have used more cameras elsewhere. In early June, a recreational fisherman who lived near Detective Bill Jarmon stopped by to offer a complaint. In the waters near the Nisqually River Delta, he had seen an aluminum boat pull up some unusually large crab pots from the deep. Commercial fishermen used circular cages as traps, which they baited and dropped to the seafloor. Crabs skittered in and scratched against the mesh but could not escape. The neighbor and a friend, in their own vessel, pulled alongside the boat and saw three large containers that held the largest Dungeness crab they'd ever seen. There were dozens of the brownish purple discs piled on top of one another, some approaching a foot long across the back.

Dungeness crab, *Cancer magister*

Commercial-scale crabbing had been illegal in the marine waters near the mouth of the Nisqually River for years. In fact, most areas in southern Puget Sound had been closed to commercial crab fishing as biologists learned more about the crustaceans' sensitivity to pollution and disturbances. Even sport fishermen could take only six a day, and these individual crabbers could no longer use dip nets or personal pots during molting season, when crabs shed their old shells to reveal the delicate new ones beneath. It was too easy to injure females carrying eggs. Eager fishermen had to wait until later in the year, when the new shells had expanded and hardened.

Jarmon's neighbor was stunned by the huge piles of the critters. He caught the eye of the big man on the boat who was clearly in charge of bringing up the pots. As the two boats passed, the big fisherman leaned over and urged Jarmon's neighbor and his friend to keep what they'd seen a secret.

Neither man did. After the neighbor spoke with Jarmon, his friend called the *Tacoma News Tribune*. The newspaper wrote a nine-paragraph story asking: Who was operating this illegal mystery boat?

Detective Bill Jarmon's mind was elsewhere a few weeks later as he wandered from the kitchen to the living room in his Tacoma home. The house sat on a rise overlooking Puget Sound and had a million-dollar view few cops could afford. Jarmon and his wife had built on the property before the region's software boom, and before a tsunami of transplants fleeing congestion from bigger cities shot Puget Sound real estate prices into the stratosphere. The Jarmons had installed huge west-facing windows in each room to capture every possible view of the sparkling water.

On the morning of June 28, 2000, Jarmon sipped his coffee and stared. This was his daily ritual. He would look across the channel below to the tip of the Key Peninsula or look north toward the Tacoma Narrows Bridge. The mile-long span offered the fastest way to reach the Sound's isolated western peninsulas without taking a ferry ride. Farther south Jarmon could see Fox Island, named for John L. Fox, a ship's doctor on the U.S. Exploring Expedition. Jarmon had lived in this area his whole life. He knew this landscape like he knew his own living room.

Jarmon had been a wildlife cop for twenty-five years. Some years he'd spent weeks at a time undercover, sizing up everyone he met. He'd once been attacked in his car by a drunk with a gun. Over time Jarmon developed a few occupational tics. One was that he noticed things—

whether he wanted to or not. He'd gotten deer poachers to confess by watching their faces and pretending he already knew their crimes, and his curiosity about an elk poacher once led him to solve a long string of burglaries.

On this clear morning, as the dawn gnawed through morning shadows, Jarmon caught a glint of sunlight hitting aluminum—a vessel cutting north, hugging the shoreline. Jarmon had always taken such pleasure in this view that his wife bought him a telescope, which sat on a tripod at the window. Jarmon trained the lens on the boat and nearly spit his coffee. He saw a thick center mast, loudspeakers, and a row of cabin windows. He was staring down the Narrows at *The Typhoon,* their informant's boat, which chugged up Hales Pass toward Fox Island, where Doug Tobin kept it moored.

Bill Jarmon after a promotion ceremony.

From its route, Jarmon presumed Tobin was headed to the marina, but it was odd that the boat was returning at dawn. Either Tobin got up ridiculously early or the vessel had been out all night. Fishermen collected geoducks south of the spot Tobin was motoring through, but it was legal to do so only during daylight hours when monitors were there to watch the harvest. On some mornings, the detective would see several geoduck boats heading home from a sunrise harvest. That could be the case with *The Typhoon,* but it would have to have wrapped up its fishing pretty damn quick.

Jarmon had been on the periphery of the DeCourville case and had worked other clam investigations, but he didn't share Volz's rich background in geoducks. In the last year, though, he'd been part of the team assigned to study the clam industry, and he'd had a chance to speak with Tobin as an informant. That had been odd. Unlike some of the other cops, Jarmon had known Tobin his whole career. Jarmon first met him in 1975—he'd caught the young fisherman on a December dawn illegally stringing a net to catch salmon. That incident had frozen Tobin in his mind as someone who bent rules. Having heard Volz grouse about the informant, Jarmon, this morning, couldn't shake his curiosity: What was he doing?

Jarmon grabbed his .40-caliber Glock, a notepad, and a radio and headed out over the Narrows Bridge. He radioed Volz, who knew geoduck rules as well as Jarmon knew the south Sound. Volz told him the only geoduck fishing open nearby was on the far side of the island and not on Tobin's route. *The Typhoon* clearly wasn't returning from a morning harvest. There could be a reasonable explanation for Tobin's travels, but Jarmon saw no harm in taking a closer look.

Jarmon raced toward the bridge that crossed over to Fox Island, reaching it just as *The Typhoon* puttered between it and a tiny sister island. He pulled over and sat back, watching through binoculars. *The Typhoon* docked between some boathouses and a sailboat. Jarmon surveyed the scene, not sure what he was looking for. He watched for a moment from his truck, then rolled in closer for a better view. He hid behind a mass of Scotch broom across from the marina. He slunk down in his seat and peered through a spotting scope.

Jarmon could see maybe a half-dozen people wandering *The Typhoon*'s deck. None looked like Doug Tobin. They offloaded several plastic garbage cans and covered them with damp burlap, then packed the cans into a wheelbarrow-style hand dolly that was so heavy it took two men to roll it up the dock to a white van. They loaded seven garbage cans into the van and packed two more in a red Ford pickup. Jarmon couldn't tell precisely what was in all those containers, but he had an idea. A man and a woman climbed into the pickup's cab and took off.

Jarmon tailed the truck as it crossed back onto the peninsula. He lagged behind, trying to remain out of sight on the rural roads, but he was low on gas and not sure how far he could go. Soon enough he'd dropped so far behind that he'd lost the truck. He pulled into a minimart to refuel and thought about the story he'd heard a few weeks earlier. Jarmon hadn't known Tobin to deal much in crab, but that certainly seemed like what he'd seen being covered in burlap. Could *The Typhoon* be the mystery boat? It was possible; there was no telling what went on in the informant's head. It also was possible that Tobin didn't know crabbing was banned, possible that there had been a special tribal season Jarmon didn't know about. *The Typhoon* also might not have been anywhere near the Nisqually. Jarmon couldn't say where the boat had last anchored.

What he knew for sure was that he didn't know that much. The cop had seen a bunch of strangers offloading what looked like crab. But unless Jarmon could prove the crab had come from the Nisqually, he might have difficulty proving there'd been any crime. The crew could have taken the crab from someplace else. And, of course, he hadn't actually seen Doug Tobin. He would have to talk it out with Volz and see what came next.

Jarmon finished fueling and looked up and saw the red pickup. It was pulling out from a strip mall across the street. He'd caught a break. Jarmon slipped in behind it as the truck headed back to the Narrows Bridge toward Tacoma. Jarmon passed the truck and noted the license number before settling in a few paces ahead. He tracked the truck with one eye on his rearview. It crossed back into the city, and Jarmon let it overtake him near the concrete behemoth of the Tacoma

Dome. He followed it off an exit and among the billboards and fast-food chains of Fife. Then the truck slowed and pulled into a small warehouse wrapped in chain-link, where someone opened a garage-style door and began unloading.

There wasn't much left for Jarmon to see. He drove back to the Fox Island marina once again and photographed whatever he could find left behind at the scene. He relayed the morning's events to Volz and wondered: What now?

~

The detectives had not paid much attention to Tobin lately, though they heard from him from time to time. After Severtson's retirement, calls came at random, offering rumors and kernels of inside information. Tobin always had an angle. He worked the detectives like a high school flirt, chatting them up and dropping tips and pumping *them* for information, too. Tobin was always feeling them out—in the fishing industry, like any business, information is currency. The cops didn't mind; it's how the game was played. They shared gossip that didn't mean much and listened, because they didn't get to choose their snitches. Plus sometimes Tobin's pointers were right on target, and a good tip was a good tip.

Just the previous fall, Tobin had offered an observation. The detectives were missing out on a lot of poaching, he'd said, and one of the illegal harvesters was someone they knew well—Brian Hodgson. Now in his sixties, the godfather of the country's geoduck industry was back on the docks and appeared to be buying and selling clams, even though he'd faced prison time, paid fines, and been banned for life from fishing in Washington. Tobin said he saw Hodgson every week on the waterfront, leaning on the hood of his truck and filling out what looked like clam-buying paperwork. The information dovetailed with something the investigators had seen themselves. While parked outside a marina that summer, Jarmon had elbowed Volz and pointed to Hodgson, who was chatting up several geoduck buyers and sellers. By the time Jarmon had followed *The Typhoon*, Harrington and another detective were officially investigating Hodgson again.

The detectives weren't sure what to make of what Jarmon had seen on Fox Island. Given their busy schedules and the lack of clarity, they passed the tip to some marine-patrol officers, who it turned out already shared Jarmon's suspicions. One of the marine officers a few days earlier had been in contact with the Squaxin Island Tribe. Someone had reported to tribal police that a crab boat working in the Nisqually region had "Typhoon" emblazoned across its stern. Those officers passed the information Jarmon had given them to tribal police. The tribal guys knew Tobin and would know if there'd been a special season for tribal fishermen. Perhaps the tribal police could sort it out themselves.

Squaxin Island stretches like a knife handle across the openings of several finger inlets on Puget Sound's southwestern shores. The island itself is only five miles long and narrow enough to walk across in ten minutes. The Squaxins, along with the Nisqually and Puyallup Indians, once dominated the southern lands of Puget Sound all the way to the foothills of the Cascade Mountains, but they signed away thousands of square miles to the United States in treaties. The tribe retained the island as its reservation lands, but tribal members soon left to make their homes on the mainland. By 1862, fewer than fifty people lived on the island. A century later, the number had dropped to four. By the year 2000, the island was empty.

Tribal members were scattered everywhere, but a small tribal community existed on the mainland not far from the island. The tribe operated a midsize casino outside the southern Puget Sound community of Shelton, which was one of the county's largest employers. It ran a gas station and trading post and a small tobacco plant that made cigarettes and cigars. It oversaw a museum and cultural center, with a totem pole out front that had been carved by Doug Tobin.

The tribe's small police force dealt with crime on Indian lands and shared some jurisdiction with local law enforcement. Tribal cops policed the tribe's portion of its fish and shellfish harvest and could cite offenders for poaching on Indian grounds. Chief Rory Gilliland was not a Squaxin Island Indian, but a professional cop and a member of the Delaware Tribe from Oklahoma. He'd been a sheriff's deputy in Olympia during the Northwest Indian fishing wars of the 1970s and had often been ordered to stop Native fishermen from catching salmon. He had seen bloodshed and gunfire and dismissed the period as "ugly." He left the sheriff's department after a long career and took the top cop job with the tribal police in 1993.

Gilliland had known the Tobin brothers for years. Gilliland liked Doug but they'd had run-ins over fishing. Gilliland had watched Doug bend the rules before and then try to talk himself out of trouble. Stopped by a tribal cop, Tobin once dialed Gilliland directly on his cell phone. "He'd play everyone to his advantage," the police chief later said. "And I considered him a friend." Tobin sometimes paid his tickets and sometimes fought them in tribal court. The Squaxin Island Tribe ran a small outfit. Gilliland knew Tobin didn't see him as a threat.

That June, the tribe had been swamped with tips about the mystery boat. Several suggested *The Typhoon* was the culprit. By the time Gilliland heard from the officers in Jarmon's shop, he'd already begun his own little inquiry. There was no tribal opening for crab, and Tobin didn't have the right license anyway. If Tobin was taking crab, it definitely was illegal, but he knew the waters better than they did. His boat could travel in zero visibility and could race from Squaxin Island in the morning to the Strait of Juan de Fuca the same night, a journey that might take a smaller boat several days. Catching him in the act might not be that easy.

Usually crab pots were marked by buoys, but just in case, a sport fisherman gave tribal police a GPS reading on a spot where he'd seen a vessel drop its pots. The tribal cops planned to wait there and see if someone came back to pick them up. Several nights passed with little action. Then one night at 4 A.M., a patrolling tribal officer spotted a boat shining a light in the Nisqually Delta. By the time the officer was able to approach, the larger vessel had switched off its lights and vanished.

Gilliland had no doubt that it had been Tobin, but the tribal force never got much more detail. They mailed Tobin a citation for fishing in closed waters, based on the information provided by witnesses. Tobin paid a small fine and that was the end of it. "My guys never really got it together," Gilliland said. By early July any buoys that might have marked crab-pot locations were gone. Whatever Tobin or his crew had done seemed over.

~

Ed Volz growled at the news. In truth, no one really knew exactly *what* Tobin had done or if he'd stopped, but it seemed clear that getting a handle on it would require more than just a bunch of cops passing around tips. If the detectives wanted to know what their informant was up to it appeared they would have to try and figure it out themselves, but it was hard to know how much time to invest. The detectives had no new leads and plenty of other cases. They weren't even sure he was still doing anything. They decided it was best to just hang back and keep watch.

In early October 2000, Tobin dangled an opportunity that made that easy: He offered to rat out another seafood dealer. Ken Li was currently driving shellfish to Canada on behalf of another broker. He and Tobin had done business before, but now the man wanted to start his own outfit and was looking to get some inexpensive stolen clams. Harrington and Jarmon headed south to meet Tobin in Fife, this time at Mitzel's restaurant, another family-style café, this one just a few hundred yards from the Poodle Dog. The restaurant sat less than a mile from Tobin's warehouse, and had become his hangout. The fisherman had grown so popular there that restaurant workers treated him like royalty. He arrived early each morning with a six-person crew and often came back later the same day. He sat at a round bench in front of the fireplace and was easily the restaurant's single biggest customer. He left hundred-dollar tips for waitresses and sometimes dropped five thousand dollars there in a month.

Tobin told the detectives that he wanted Rory Gilliland with the Squaxin Island Police to be present if Tobin sold Li illegal clams during a sting. The detectives had no objection. They saw Li as a minor bust but thought it might lead to some insights. It might help them learn how geoducks moved through Canada—and it would get them inside Tobin's plant.

At 2:30 P.M., they followed Tobin down a short dead-end road to his warehouse. The place was made of concrete and stainless steel and was filled with tanks, tubs, and refrigeration units, an unspectacular place that smelled vaguely of crab. Tobin would work the sting but demanded the cops reimburse him for any clams he used. They might exchange several thousand dollars' worth, and Doug knew the detectives would confiscate them at Li's arrest. The detectives stayed noncommittal but agreed to address it later. Finally Tobin called Ken Li while Harrington listened in. "You give me the product and I'll pay you cash," Li told Tobin. "Don't worry. I've done this many times in the past."

The next afternoon, Gilliland boxed clams in the warehouse, pretending to be just another of Tobin's workers. Jarmon crawled into the loft where he could watch. The hard plywood floors put his legs to sleep, and he could barely hear over the whir of a nearby generator. But from here he'd be able to see Li and Tobin. This was merely a dress rehearsal. The two men were only meeting to make plans.

Tobin opened the warehouse doors for Li's green Mazda minivan. Li was well groomed and graying in a denim shirt. He looked like a young grandfather on his way to play golf. He helped Tobin unload several empty milk crates near a set of holding tanks. He complained that during their last transaction Tobin had shorted him fifty pounds. Tobin dismissed the weight difference as an oversight and promised to make it up this time. He joked that Li's minivan looked like something an old woman would drive. "It's great," Tobin said. "No cop would ever follow you in that." Li laughed and agreed it was very discreet. Then he made Tobin promise not to tell anyone else he was branching out. He was pocketing nearly five thousand dollars a week, and he showed Tobin a wallet stuffed with one-hundred-dollar bills. When Jarmon crawled down after Li left, Tobin could barely contain his excitement. He was clearly pleased with the way things went.

Two weeks later, the detectives set up their sting. Li made his illegal purchase as the cops looked on, then drove off toward Interstate 5. Tobin had insisted the detectives make a show of arresting him, too. He'd said he didn't want it to be obvious that he'd turned in Li. The cops slapped handcuffs on Tobin and put him in a car and drove him to where patrol officers had pulled Li over down the street. With Tobin looking on from inside the car, the cops searched Li's van and found several thousand dollars' worth of geoduck hidden beneath a tarp. They also discovered a little surprise: a quart bag stuffed with marijuana. Li's eyes widened with concern: "Not mine! Not mine!"

The detectives didn't really doubt him. Li seemed far too panicked to be lying, and the cops had other suspicions anyway. They had never known their informant to use drugs, but they knew weed was common among his fishing buddies. Tobin had asked to be present during Li's arrest—an unusual request, even for him. The detectives figured Tobin tossed the dope in the van just to see what would happen next. Volz wondered if Tobin had really wanted to see Li's arrest just so he could watch the seafood dealer freak out.

They didn't say anything about their suspicions. The cops didn't want Tobin to know they were paying attention. They didn't want him to suspect they were watching him that closely. They needed him to think he was smarter than a bunch of fish cops.

By early spring 2001, word filtered back: Tobin was out there poaching crab for sure. Volz was too tired to be that angry. He'd hesitated to work with Tobin in the 1990s in part because Dave Ferguson had been so draining. But Volz had only kept tabs on Ferguson for a year. He'd now dealt off and on with Tobin for almost five, and no one else in law enforcement felt any responsibility for him. Volz discussed the matter with Detective Kevin Harrington. They agreed it was time to take Tobin seriously. No other agency had the means or inclination to stop him. Given how frustrated they'd been about the DeCourville case, any investigation might look like a vendetta. The detectives talked it over with their colleague Bill Jarmon and agreed that he should lead the probe. Jarmon had played only a cursory role in the DeCourville cases. No one could suggest he had an ax to grind. Plus Jarmon had shown interest the previous summer when he chased *The Typhoon* all the way to Fox Island. Once set in motion, Jarmon was relentless. He wouldn't stop until he found out everything.

But what exactly was there to find? They didn't know much at all, only that Tobin was taking crab from somewhere. Where, and how much, was anyone's guess. They had whispers. Chasing Tobin would be like grabbing at smoke.

In early March, the detectives attempted to spy on *The Typhoon* and its captain, but they had little to go on. Jarmon slipped out at night from his home in Tacoma and watched through binoculars as Tobin and his crew loaded the boat off Fox Island. Another detective returned a few nights later and did the same. A tipster told Ed Volz he'd confronted Doug, who'd claimed he was working on an undercover project with the Feds. But this man was certain that Tobin was just stealing and keeping his efforts quiet by threatening other fishermen, promising to make them targets of *his* investigations if they made a stink about his activities. It was laughable, but actually pretty smart, Volz thought. Only Volz and the other cops knew how crazy that would be; Tobin hadn't worked with federal agents in years. But fishermen wouldn't know what to think. Many didn't dare ignore his claims.

Investigators watched from the bushes on five more nights in March and April as Tobin and his crew shoved off in the darkness. Sometimes the detectives drove back to the mainland, where they could see the boat anchored near the Nisqually. Other times they were fairly certain they saw the boat offloading geoducks. No doubt, the activity was suspicious. But while fishing for crab in a closed area was against the law—as was fishing for geoduck after dark when monitors weren't watching—boating at night was perfectly legal. What exactly was Tobin into, really? They still needed to know a whole lot more.

Their first significant break came with another whisper, a tip from a crab fisherman who lived along the coast. The fisherman had spent a lifetime in the salmon capital of Westport and for years had been a source for Volz. He said a woman he knew, Heidi Mills, had fished illegally with Doug Tobin but was tired of fighting him over pay. The crabber urged Volz to reach out to her.

Heidi Mills and her boyfriend met the cops in Olympia in late April 2001. She was a thin chain-smoker in her late thirties who had spent more than a decade in commercial fishing. She told detectives she'd rented plant space from Tobin a year earlier planning to pack and ship her own crab. But Tobin had also hired her and her boyfriend to work on *The Typhoon*. Heidi told the detectives that over time she'd figured out that Doug regularly broke the law, not just with crab but also with geoducks. He didn't bother much with any kind of paperwork and regularly went out fishing at night. When she'd confronted Tobin about the lack of documentation on one shipment, he'd tried to tell her it didn't matter because the load was bound for Canada. He swore to her that at night he only fished for crab—not clams—and did so only to keep fishermen from trying to steal his gear. But Heidi hadn't just fallen off the salmon boat. She'd managed the line on a factory trawler in Alaska, had spent time in the wholesale lobster business, and had worked on crab boats on the Pacific. She understood the law and had known plenty of cheaters. Tobin apparently hadn't even thought through the lie he'd told her; he'd even taken her boyfriend out night fishing. Her boyfriend had driven the red Ford pickup the morning Jarmon had followed from Fox Island. "You're poaching," Heidi said she'd told Doug. Tobin had just shrugged and snapped: "Call it what you want."

The detectives wished Heidi's story weren't true, but they believed it. It was what they'd suspected all along. Their informant didn't just bend a few rules. He twisted the law he'd spent years helping them enforce. To the wildlife cops the message couldn't have been clearer. Their informant had become the equivalent of a rogue spy who no longer answered to anyone but himself. He did as he pleased, and thought he was untouchable. It wasn't as if law enforcement had done much to convince him otherwise.

During their interview, Heidi and Detective Kevin Harrington frequently stepped outside to smoke. During his academy training years earlier, instructors had urged Harrington to keep witnesses off balance by withholding cigarettes. But as a chain-smoker, Harrington knew how he'd respond: He'd be pissed. He took a different tack. When he wanted information from someone who shared his habit, he usually found a way to join them for a smoke. It gave him a chance to connect in a looser setting, and witnesses usually let their guard down.

"He said he was working with the Feds," Heidi told Harrington after one such break. "I didn't know if he was blowing smoke or what, but that's what he said." And it seemed to her that Tobin was pretty good. At night he used radar and night-vision equipment to ensure no one else was on the water. He mixed clams he collected legally in with the ones he poached, which made it much harder to tell what had been stolen. He'd found an elegant way to smuggle. When a business shipped seafood to or from the United States, laws required paperwork that told authorities the shellfish were obtained legally. Tobin's approach was low-tech and low-risk. He doctored the paperwork and put the clams in a van and simply drove them across the border. Once in Canada someone else would worry about Canadian rules and figure out how to fly the geoducks to Asia. Tobin also shipped clams by air to California and often doctored those records, too.

Heidi didn't like what she'd been doing but figured to get paid at all she had to keep working. When she complained about the situation to Tobin's brother, he just shook his head and said Doug was begging to get caught. But still her money never seemed to come. Heidi even recalled a night when Tobin had gone out with her boyfriend and illegally collected several hundred pounds of geoduck. Tobin had refused to pay him because he said the geoducks were to be used in an undercover sting.

Harrington thought about what that meant. Tobin had *stolen* the clams he'd used during the Ken Li sting. And not only had he demanded reimbursement from the cops, he'd also ripped off the fishermen who helped him get them. Harrington had to admit: The guy had balls.

Frustration mounted that summer as the detectives struggled to grasp the depths of Tobin's deceit. Through May, June, and July, they gathered some new evidence but nothing that would break their case wide open. Then a new lead came to them. Tobin had forgotten the key lesson that had led to Brian Hodgson's downfall—take care of the people working for you. Sooner or later aggrieved employees strike back, and one of them might have enough to hang you.

On a late summer day in 2001, one of Tobin's workers, Keith Smith, appeared out of the blue and stood at the front counter of the Washington Department of Fish and Wildlife headquarters in Olympia. He was stout, with a shaggy brown mop, sunken eyes, and weight lifter's shoulders. He told the receptionist he knew about a crab-poaching operation. No cops were around, but the receptionist offered to take his number and have a detective get back to him. Smith said he lacked a telephone but left his name and address and said he would call back. He never did.

The message worked its way to Detective Jarmon. Another officer in the department might have tossed it in the trash—it was a random tip with no details from a man named Smith who had no phone—but Jarmon didn't think like other cops. When he saw that the tip involved crab, he headed through the lowlands and back roads and into the woods, trying to read the address Smith left in his scrawled hand-writing. Early one evening in late August, after several wrong turns, Jarmon pulled into the forested driveway of a tiny rental home along a road connecting southern Puget Sound and the coast. He stepped from his Expedition and introduced himself to Smith.

Smith and Tobin had met three years earlier when Tobin started carving his welcome pole for the Port of Olympia. Smith worked in the warehouse next door, and Tobin wandered in, asking if he could drop his shavings in the Dumpster. The men chatted, and within weeks Smith was working on the pole. Tobin talked the whole time about the great money they could make fishing and quickly convinced Smith to join him aboard *The Typhoon*. The arrangement worked for a while, until Smith felt cheated on pay.

Smith knew he was in trouble. He had worked with Tobin when Tobin broke the law. He apologized to Jarmon for ever getting involved. When Jarmon asked why he bothered to come forward, Smith gave an answer that caught the cop off guard. Smith was angry that Tobin had cheated him, but he also said he felt guilty about what he'd done. He wanted to do something to put an end to it. Jarmon tended not to buy such lines, but Smith struck him as sincere.

As an informant Keith Smith was no Doug Tobin but he had some-thing the cops could use. He'd come to know Tobin's operation inside out. He knew where Tobin worked and with whom he fished. He could identify locations on a map. In a series of taped interviews over the next several weeks, Smith outlined everything he knew and provided the names of nearly everyone who had ever worked on *The Typhoon*. What Smith described was far larger than the cops had thought: Just before midnight three to four nights a week, sometimes five, *The Typhoon* slipped away from the Fox Island boat ramp. Some nights were crab nights, others were for geoducks. With Tobin at the helm, the boat headed for clam-rich destinations, areas Doug knew from survey maps, or places he'd explored using underwater video. Once in place, Tobin ordered all lights extinguished and sent a crew member to the stern with night-vision goggles to scan the horizon for oncoming vessels. No one could speak, and smoking was usually banned; another boat or someone on shore might see the glow.

Smith was a deck boss who oversaw the other workers. On geoduck nights, he tended to the divers below. He helped them dress and enter the water and managed their hoses while they pulled clams off the floor. He stacked geoducks in cages and assisted divers out of the drink. A rotating crew of more than two dozen men and women worked on or around *The Typhoon,* many known to one another only by nicknames. Spook. Hollywood. Slim.

What Smith described sounded like a circus. One of the best divers was afraid of deep water and panicked at the sight of floating logs. Another lived in mortal fear of sharks. A third sometimes surfaced with his hoses snarled in a dangerous rat's nest. And he was among the more safety conscious. When one diver's oxygen ran out sixty feet down, the worker monitoring the air compressor didn't even notice. Another crewman sprinted from the cabin in such a rush to flip on the reserve air supply that he tripped and broke his leg. The high rollers in the crew worked geoduck rather than crab. An average diver like Slim, a six-foot-six twenty-six-year-old who weighed less than two hundred pounds, made $500 to $900 a night. The best diver sometimes pulled down two thousand.

Everyone understood the house rules. If the cops arrived, they were supposed to just dump the stash and run, and not slow down or stop unless someone pulled a gun. If someone was working down below, the crew members would use a knife and slice the line to his water jet, the stinger. That way they could say they merely had a diver down exploring. Clams would remain stacked on the side of the boat where they could easily be jettisoned in a pinch.

The Typhoon did most of the hard labor, but workers sometimes towed a small skiff or a Boston whaler to transport the geoduck to a nearby marina. That way *The Typhoon* could dock at Fox Island empty. The operation had been going on so long and went so smoothly that Tobin often didn't bother going out anymore himself.

The crab operation was easy but less lucrative, Smith explained, usually netting each worker three hundred dollars or less a night. Most crabbers tied their pots to buoys so anyone floating by could tell what rested below. *The Typhoon*'s crew was more devious. The men strung pots together with a cable called a ground line and dropped everything to the seafloor. The crew would take a GPS reading and return to the same spot days later with a grappling hook. They would snare the ground line and retrieve the pots. On crab nights, Smith baited, lowered, and retrieved the gear. Except for the very moment they pulled the pots aboard, no one would ever know what they were doing.

Tobin had found the crab mother lode at the Nisqually Reach by hiring an aging poacher who had fished there in the 1970s. This gentleman's sole job on *The Typhoon* was employing his vast knowledge of the south Sound to locate deepwater stores of monster crab. The crabs he found were, in fact, huge—about 30 percent larger than most in Puget Sound. They had gotten that way because commercial crabbing had been banned in that spot for decades. Tobin worked several hundred crab pots at one time, most in waters more than 150 feet deep. The crew collected fifteen hundred pounds every three days. Smith said he'd helped Tobin collect those crabs at least two hundred times in about three years. He estimated he'd poached geoduck 150 times.

Jarmon did some back-of-the-napkin arithmetic. If his math was right, Doug Tobin had taken hundreds of thousands of pounds of geoduck and clam. That would make the informant the biggest wildlife thief they had encountered in two decades. Even if he'd only poached since meeting Smith, the detective figured their informant had made off with $3 million in stolen shellfish. And Tobin had learned his tricks working with cops.

Jarmon dialed Detective Volz.

"I wish I could say that I was more surprised," Volz said.

The magnitude of what Tobin had done sunk in slowly. He'd worked his fishing connections to generate on-the-money tips. When he fed that information to authorities he must have felt reasonably sure they weren't watching him. The more he pointed to everyone else, the less likely it would be that cops even gave him any thought. Whether by luck or design, Tobin had also helped the cops clear away his chief competitors. When DeCourville went to prison, Tobin began selling to his customers. Getting rid of Ken Li probably helped him, too.

But Tobin's brazenness fueled new tips. Several weeks later, Detective Jarmon got wind of an unusual telephone call from a community college English instructor on the Key Peninsula. The instructor lived within view of Wyckoff Shoal and complained about seeing a boat anchored in rough weather without lights at night. When he looked in the morning, the boat would be gone. Intrigued, this instructor began paying closer attention. At night, he watched out his bedroom window. Sometimes when he saw the boat, a flashlight on deck would flip on briefly. Then just as quickly the light would blink out. He knew boating at night without lights violated Coast Guard rules. He presumed that meant someone was trying hard not to be seen. He saw shadows in moonlight of men wandering around the deck.

He started writing what he saw on three-by-five note cards. More than a dozen times in six months he recorded his observations. One night the boat turned so the cabin faced his home. The instructor could see faint yellow and orange light slivers leaking around cabin window rims. Someone had covered the panes with construction paper or dark curtains. One morning he rose before the sun and heard the motor cough; he watched through predawn light as the boat took off. At a friend's suggestion, the instructor called wildlife officers. Jarmon convinced him to mail in his notes.

The detectives knew catching Tobin would require extraordinary measures. The detectives couldn't follow him by boat. As tribal police had found out, *The Typhoon* was fast and well equipped. A patrol boat probably couldn't get close. The only option would be to tail the boat from land. But at least now they had eyes and ears inside. The detectives had asked Heidi Mills and Keith Smith to remain in touch with Tobin. Smith in particular steadily fed Jarmon information. The detectives now stood a chance of at least learning when *The Typhoon* was going out. They might even find out where it was heading.

Several times through late October and early November, Jarmon, Volz, and another detective named Charlie Pudwill slipped out after dark when they heard *The Typhoon* might be out. Sometimes the detectives arrived after *The Typhoon* left its moorage. A few times they set up video cameras in the bushes outside the marina, waiting to record what happened on the boat's return. The detectives spent the entire night of October 25 camped in their truck among the weeds. When the boat returned at dawn, the crew stepped off but carried nothing. Five nights later, through high winds and heavy rain, the detectives watched several men wander around the deck of *The Typhoon*. They waited for five hours, but the boat never left the dock. Night after night, surveillance continued with little change. At dawn, the video camera would catch several men and women walking off the boat without contraband. Rarely was Doug Tobin even among them.

On November 13, their luck turned. Just after midnight the detectives trailed the boat from a shaded bluff and chased it on foot during its journey through Drayton Passage until it finally disappeared around Devil's Head point. Moments before Volz and Jarmon had decided to give up, Volz had seen the boat jet back. Then Detective Pudwill called from his station near Wyckoff Shoal. *The Typhoon* was back, idling across the channel. Volz and Jarmon were inside Jarmon's Expedition, bloody from scraping through blackberry brambles and damp from wandering the beach in the rain. They were exhausted but agreed to meet Pudwill. If Tobin's crew was ready to work, then so were they.

chapter eleven

THE HUNT, REDUX

November 13, 2001

From a pea-gravel beach a few minutes later, the three detectives watched the boat sputter around the channel, then cut its engines and lights. The men could see its dim outline across the channel through the predawn haze. In the distance, they could hear the air compressor throb and whir. It was after 3 A.M., and the detectives had been working since the previous morning. Now another diver had finally been sent below. It appeared their night wasn't over after all.

For an hour and a half, the detectives kept the boat in view. Pudwill propped his spyglass on a tripod, and Volz told him to lock the scope in position and not move it, even if the boat peeled away. Later, they would need to know precisely where the boat had idled. Volz and Jarmon hiked back to the Ford Expedition. They would return to the dock on Fox Island, pull in behind the bushes, and try to catch the crew unloading. Later they'd take a boat back to Wyckoff Shoal and, using video of the seafloor, see if they could prove how much the crew had taken. They hoped, just once, to see Doug Tobin's face.

Morning light filled the sky by the time Volz and Jarmon crossed the bridge onto Fox Island. *The Typhoon* was already moored at the dock where the crew was offloading the evening's take. A man and a woman with a blond ponytail lugged milk crates from the boat. They stacked the containers in a parking lot, where another man used a hand dolly to cart them to a van. From the bushes, the detectives saw wet geoducks piled end to end in the crates. They watched the crew carry off *The Typhoon*'s gear. The woman and four men quickly cleaned up, moved equipment to three cars, and said their good-byes. The detectives sat until the vehicles drove off. Where the hell was *The Typhoon*'s skipper?

Volz and Jarmon could have jumped out like drug cops hassling street-corner dealers. They could have arrested those coming off the boat. But a single night of geoduck poaching would only land the poachers a puny theft charge. Most likely they would wind up with a small fine or probation, and Tobin would just change tactics and keep going.

There was little to do but press on. Volz called biologist and shellfish diving expert Don Rothaus. They agreed to meet in three hours at another marina on the peninsula. It would take Rothaus and his crew that long to prepare the agency's boat, which gave Volz and Jarmon enough time to catch a catnap. Jarmon pulled the SUV over, and the two men settled in for some shut-eye. It would be their first in more than twenty-four hours.

Later that morning the research vessel *Clamdestine* sliced through icy waters toward Drayton Passage. The day was overcast, but visibility was good. Rothaus had piloted the boat for fifteen years conducting underwater surveys of Puget Sound's abalone, sea cucumbers, urchins, scallops, and geoducks. The twenty-eight-footer was a thick-hulled fiberglass boat and had been constructed to take body blows. It had been bounced off rocky outcroppings and slammed by fifteen-foot waves. The large working deck was roomy, ideal for ferrying guests. Volz and Jarmon stayed outside in the chill to keep awake. They had worked intermittently with Rothaus for years. They didn't go into detail about what they were doing. They told him only that they wanted any evidence of digging on the bottom. Rothaus mostly used the boat for research. Only occasionally did he get to use it for law enforcement, and the chance to take part in the investigation clearly animated the biologist. He was talkative and pointed out the irony: a boat named *Clamdestine* taking part in espionage work involving clams. Rothaus asked if he needed a code name, and the detectives jokingly called him "The Weasel." Rothaus didn't seem to mind.

Rothaus maneuvered the boat until he heard from Pudwill that the *Clamdestine* filled his scope. Pudwill had shivered onshore with no food, coffee, or breaks for nearly ten hours. They dropped anchor where Rothaus and another biologist planned to dive and sweep a video camera across the seafloor. Only if Rothaus could prove clams were missing from the channel and had been removed recently could detectives argue that the crates of geoducks offloaded from *The Typhoon* came from that spot and nowhere else. At 11:20 A.M., Rothaus and fellow diver Michael Ulrich slid into the water with oxygen cylinders on their backs and carrying underwater lights. They settled to the bottom thirty-one feet below. Fifty minutes later they emerged and Rothaus grinned. He and Ulrich climbed into the boat. Volz asked to watch the recording right away.

The detectives hovered around the portable display screen. Down below, the divers had both headed west, Rothaus holding the camera. Within minutes, the detectives saw what Rothaus now described as a long row of dig holes in the muck from which divers had yanked clams. Rothaus had filmed through the green water as Ulrich probed the compact surface around a depression in the sand, showing how solid the floor was. Then Ulrich stuck a probe in the depression until it sunk sixteen inches—something clearly had been removed. He moved a few feet and found another hole two feet deep.

For twenty-five minutes, Volz and Jarmon watched row after row of these divots appear on the tape. Given the currents, loose sand would have compacted after only a few tidal cycles; the presence of the holes meant they were fresh. The two divers also showed them furrows in the sand created by the drag of a fisherman's air and water hose; those furrows would also not last the passing of a few tides. Furthermore, hundreds of varieties of thin white tubeworms usually live among the sand below the surface. As divers pluck geoducks from the muck, these worms get kicked from the mud and land on the seafloor, a telltale sign of clam harvests. Rothaus pointed out that worms now littered the surface. He pointed to discolored and discarded geoduck shells sitting on the ocean floor being nibbled on by purple-legged crab, starry flounder, and flatfish.

For the first time, Volz and Jarmon had proof *The Typhoon* had stolen several thousand dollars' worth of clams. The detectives had documented every step of the operation, from the shore to the water to the seafloor and back to shore. The biologist was beaming, proud of his own role. But Volz couldn't bring himself to return the smile.

Rothaus pulled the anchor and fired up the *Clamdestine*. Volz and Jarmon stayed in the cabin. Along with Pudwill, the detectives had spent more than a hundred hours apiece in the last two weeks tracking *The Typhoon*. It was now sixteen months since Jarmon had seen Tobin's boat from his living room. They had finally witnessed men and women stealing clams, perhaps as much as two thousand pounds' worth in one night. But they still lacked what they wanted most.

To build a case against the region's most prolific wildlife smuggler in nearly twenty years, they had to directly tie him to the crimes. If they wanted to charge him with conspiracy, racketeering, felony theft—real crimes—they had to make the case airtight. They had to do more than link *The Typhoon* and its crew. If they were to shut Tobin down for good, they had to actually catch *him* in the act, and more than once. Yet despite everything they'd seen through a long, exhausting night, the skipper of *The Typhoon* had been nowhere in sight.

The detectives regrouped. After a few days to catch up on sleep, they decided to expand the investigation's scope. If Tobin no longer drove the boat himself, they would have to tie him in some other way. So far, they could prove a boat registered to Tobin took geoducks at night and that employees on his payroll drove off with stolen shellfish in his van. Getting him on videotape demonstrating knowledge of those events could seal his fate. They needed to be able to make it clear to a jury where Tobin fit in.

Less than a week after their epic all-night chase, the same three cops conducted all-night surveillance one more time. This time, they followed the geoduck-filled van to Fife, where it pulled into Tobin's seafood plant. Two nights later, they followed his crew again. Detectives Volz and Jarmon even tailed the buyer who took the load to the airport. Outside Alaska Airlines' cargo docks on November 21, they watched the delivery man unload white cardboard boxes. The detectives went to the office and collected copies of the shipping documents and the bill.

They were still at Sea-Tac when Jarmon's phone rang just after noon. Pudwill was calling from his perch outside Tobin's processing plant. He'd been kneeling at the window in an unmarked, state-owned motor home, shooting undercover footage with an eight-millimeter camera. Through the blinds and over a fence cloaked in blackberry brambles, he'd watched one van unload and another pack up for the airport. No one shut the garage bay doors, so Pudwill just kept filming. He saw people spraying down empty geoduck crates and cleaning up from a night of work. Finally, around lunchtime, Doug Tobin wandered outside wearing a thick North Face jacket. He handed a wad of cash to one of his workers and said something that made the others laugh. Then, for a brief moment, he looked around, a merry fish king surveying his empire. For perhaps fifteen seconds, Tobin stared off toward Pudwill. Pudwill didn't know what he was doing but knew better than to worry that he'd been made. He'd chosen his position carefully and trusted his instincts. There was no way Tobin could know he was being watched. As quickly as he'd turned, Tobin looked away. Pudwill called Jarmon to say he'd finally gotten the money shot: Doug Tobin had been present and in charge during the commission of several crimes.

A few days later Tobin paged Kevin Harrington. The informant chatted for a while, then asked when he'd be reimbursed for the geoducks he'd lost during the Ken Li sting. Harrington promised to check into it. The men talked a bit longer. Tobin said he had heard that Harrington was looking at Brian Hodgson and his partner. Tobin said he could help them bring down the partner. He told Harrington he also had information about a cheating geoduck company in Canada—the business he'd worked with for five years, the one owned by Julian Ng and Jeff Albulet. He could deliver it "on a silver platter."

Finally Tobin got to his real point. One of his crew members had heard that fish cops were snooping around *The Typhoon:* What was that about?

Harrington said he didn't know.

Tobin said it took place when the boat was moored off Fox Island. What were Harrington and Volz doing?

Harrington stayed vague, and they agreed to get coffee the following week.

Five days later, Tobin called Volz after not speaking to him for nearly six months. Volz heard something different in his voice. Fear? Tobin asked Volz what he was working on, but Volz was elliptical. Again, Tobin volunteered to give up the Canadians and said there was more poaching going on than ever.

"But I thought you and the Canadians were partners," Volz said.

"Not anymore."

Volz asked Tobin if he was gathering any geoduck.

"A little here and there. Not much."

Tobin was probing. He'd tried casualness with Harrington and directness with Volz, but it was clear that neither offered satisfaction. The next day, Tobin again spoke with Harrington and asked him to track down Special Agent Richard Severtson. Tobin said he was furious that no one had told him Nichols DeCourville had finished his prison term. He feared for his safety. Oh, and the Feds still owed him money for the gasoline, cell phones, and geoducks he'd used during a year of work undercover. Perhaps it was time he sicced his attorneys on the U.S. government. He complained that people were spreading false rumors about him, saying that he was out there "doing this or doing that." The same rumors had trailed him when he'd worked for the Feds. Now he was getting threatening calls—payback for helping the cops. Someone with a heavy Italian accent, probably DeCourville, had called to tell him his days were numbered.

Harrington simply waited Tobin out. He suggested they meet at Mitzel's in Fife for a chat, and Tobin agreed.

Volz and Harrington drove down the next week, but Tobin never showed.

———

In January and February, Tobin was still calling the detectives and questioning everyone around him. Maybe he knew, or maybe he sensed, that things were closing in. His paranoia rattled those around him. In early February, Heidi Mills's boyfriend left a message on Harrington's answering machine: "Sir, if you guys do anything to jeopardize Heidi with this Tobin case, I will fuck you to the end of the earth. I'll blow your case against Doug Tobin right out of the water. So you guys better make sure that you have her best interests at heart," he fumed. "Other than that, have a great day."

Then the cops almost blew up the case themselves. Well after midnight in early February, a pair of patrolling game agents came across parked vans and an empty boat trailer near the docks on Fox Island. The patrol officers thought someone might be collecting clams illegally and waited for the boat to return. Around 2 A.M., two crewmen came ashore with the small skiff, and the two officers approached and searched the little whaler. When they found nothing but a handful of old mussels knocking around the boat's floor, they let the men go.

Immediately after hearing about the incident, Volz dialed his command and let loose a tirade. Patrol officers had been told to stay away. They had no business poking around the detectives' investigation. They might just have screwed up everything.

Tobin called Harrington's car phone a few mornings later. He started out cagey and didn't mention the incident, but managed to suggest that someone was setting him up—possibly DeCourville. Then he mentioned Ken Li and a fellow tribal member Tobin said had once asked him to dive illegally. He acknowledged there were stories floating around about him poaching, even about his involvement in money laundering and racketeering. But the truth was, Tobin said, that he'd started the buzz himself. The Feds had instructed him to, as a diversion. Now friends told Tobin they saw him being followed.

"I know it's not fair to ask," Tobin finally said, "but . . . do you have me under surveillance?"

"Not that I'm aware of," Harrington said. He tried to sound bored and slightly amused. "This is the first I've heard about it."

Tobin started making excuses. He told Harrington he'd made a game where his crew would go out for several nights and pretend they were poaching. Sometime after midnight, he'd ordered two crew members to sit in a skiff near Drayton Passage. Tobin wanted to see if anyone would notice. Sure enough, officers were waiting when they got to shore. But it wasn't even possible for them to have been harvesting geoducks from the skiff, he said. It wasn't big enough to hold diving equipment! Then again, having his employees "fake poaching" was perhaps a dumb idea, he admitted. He asked Harrington to find out what he could and call him back.

Harrington called back that afternoon as another detective listened in. Harrington reassured Tobin that the officers who stopped his boat had merely been conducting a routine shellfish-poaching patrol. It hadn't been part of any surveillance. Just a couple of rookies checking stuff out.

"That's not what they told my boys," Tobin said. "They told my boys they'd been watching for a while."

"That's what we tell them to say. Sometimes when they do, people confess to all kinds of things."

Tobin snorted and said he guessed that made sense.

Tobin could not have known how many of his colleagues had already turned against him. Several informants now fed detectives information. In addition to Heidi Mills and Keith Smith, Tobin's crab fishing contact in Westport had finally had enough. He felt Tobin owed him eighteen thousand dollars for a deal he'd made with Doug for some crab pots. Increasingly frustrated with Tobin's refusal to pay, the man started calling detectives after every conversation with Tobin.

Crab fishing is the country's most dangerous profession, and this contact had been engaged in the deadliest kind—fishing for Dungeness crab off the coast of Washington. Fishing the protected waters of Puget Sound is nothing like working the open waves of the Pacific Ocean, where crabbers fish through early winter and where squalls turn seas their roughest. The Seattle-based fishermen who ply the rough waters off Alaska's North Pacific for king crab get all the attention. Soon they would be featured in the Discovery Channel hit *The Deadliest Catch*. But for sheer statistical probability of death, Alaska's king crabbers didn't come close to matching the risks of Dungeness crab fishing off Washington and Oregon. Powered by cowboys and mavericks, the small fleets regularly lose crew members to falls, see boats crushed against hidden sandbars, and watch entire rigs capsize in ripping crosscurrents. Dungeness crabbers die twice as often as Alaskan crabbers.

The contact was a cool customer. He'd done some business with Doug Tobin and spent time touring *The Typhoon*. Finally he paid a visit to Volz's north Seattle office. He confirmed information the cops had from other moles and told them Tobin was unaware that his crew was imploding. Tobin had become so tightfisted that the mother of one of his geoduck poachers actually had threatened to sue him for not paying her son. This informant agreed he would keep visiting Tobin and share the information he gathered with Volz.

In mid-February Tobin called Harrington again. The phone calls had become tiny quick-burst chess matches, as Tobin tried to manipulate their relationship. He was still waiting for his money from the Ken Li sting. Harrington said he was working on it.

"Rumors are flying about me poaching," Tobin said. "I hear my tribe might even pull my quota."

Tobin asked again for Severtson's number. He said he was still angry that no one let him know DeCourville was out of prison. But Harrington said he shouldn't blame Severtson, who was done working and probably not even aware of DeCourville's release. Tobin dialed it down. He wanted Harrington to know he didn't really blame anyone.

"You could come down here right now and slap handcuffs on me and I wouldn't take it personally," Tobin said.

Harrington wondered if Tobin knew what was coming.

Volz's crab fisherman called a few days later to say he'd had coffee with Tobin, who wondered aloud why so many of his old employees had disappeared. He also said he now had 250 crab pots and that he was fishing more than four hundred feet down. Tobin was looking to buy 200 more, and if he got all of them going at once he said he might take a break from chasing clams.

Early in March, Adrian Lugo called his former partner to say he was concerned. Despite dissolving his business relationship with Tobin five years earlier, Lugo still felt empathy for him. A few years earlier, one of the last times they'd spoken, Lugo had put Tobin in touch with a developer from the Seattle suburbs. Lugo had been in a discussion with the developer about fine art. In the guy's office hung a gallery of Native American carvings: Salish paddles and tribal masks, carved wooden fish, and cedar totems. He showed Lugo a book of photographs of all his other pieces. Lugo recognized one as Tobin's work. When Lugo mentioned Tobin's name and that they were friends, the developer's eyes had widened. He would eagerly hire Tobin to carve more pieces, and he particularly wanted a sixteen-foot canoe. He said he would pay up to seventy-five thousand dollars. The developer assured Lugo he knew others who also would buy Tobin's art. Lugo was excited for his old friend. He introduced the two men, then stepped out of the picture.

Lugo felt he'd presented Tobin with a promising opportunity, but now he was hearing disturbing stories. He invited Tobin to come see him in his office. There, Lugo sat his old friend down and asked about the rumors. A mutual friend had told Lugo that Tobin was stealing shellfish. Lugo wanted to know: Was it true?

Tobin's response caught Lugo off guard. He put a finger to his lips and quietly got up and closed the office door. He told Lugo he was working a big case for the Feds as an undercover agent. He couldn't say anything more about it. He asked Lugo to trust him—it was all aboveboard.

Lugo admitted he was skeptical. No way would the Feds encourage Tobin to dive illegally, certainly not at night. He reminded Tobin that art buyers would still pay him tens of thousands of dollars for his carvings. Lugo had already made the arrangements. Doug would just have to follow through.

"Don't ruin this," Lugo pleaded.

Tobin assured Lugo that he needn't worry. He wasn't doing anything wrong.

Lugo let it drop, and the men parted ways.

On March 13, 2002, Tobin called Harrington one last time and begged for a meeting. He had information about a huge smuggling operation.

He didn't want to talk about it on the phone but asked if Harrington could meet him in a few days.

Harrington had been expecting this call. The detectives had discussed just what he would say. Jarmon had given Harrington an idea. Harrington had built up his rapport with Tobin over years. He'd shared numerous gripes, usually about his superiors. This time Harrington would put on a show. Harrington told Tobin he would really like to get together, but he'd be stuck in meetings at the U.S. attorney's office over the next two days. And that wasn't all.

Harrington worked himself into lather. After that, he said, he would be out of town for at least a week. Despite being buried with his own poaching cases, the bureaucrats who ran his department had other ideas. In the middle of one of his busiest periods, they were pulling Harrington off for mandatory training in eastern Washington. Thirty years on the job and he still needed training?

That wasn't even the craziest part, Harrington confided. He told Tobin he wouldn't believe what came next. Not a single cop was going to be around the office or on the water all week. The people in charge won't let any of us stick around.

Harrington spit his final words in disgust: "There won't be anyone watching Puget Sound!"

chapter twelve

THE WHOLE WEST COAST

Detective Kevin Harrington's gambit paid off: *The Typhoon* went out poaching that very week.

Between Doug Tobin's calls and information gleaned from tipsters, the detectives were now more in sync with *The Typhoon*'s maneuvering than Tobin's own team. On March 17, Detective Bill Jarmon called Heidi Mills to ask when the crew would pack the geoducks poached the night before. Heidi told him the divers had not actually gone out. Jarmon quietly assured her they had. She called back fifteen minutes later, mildly amused. She'd been wrong. Workers were boxing up the merchandise as they spoke.

Detective Ed Volz drafted detailed search- and arrest-warrant affidavits, and the cops prepared a plan of attack. Detectives, patrol officers, and cops from other agencies gathered that night in a Gig Harbor motel. No one would be allowed to leave the building until morning. Gig Harbor wasn't a large town; the detectives didn't want someone to overhear cops chatting about their assignments over chicken fried steak at Denny's. From inside an undercover motor home, Coast Guard officers and others watched *The Typhoon* tie up at a dock. They would wait to take Tobin and his team the next morning. Some of Doug's crew carried guns.

At 6:30 A.M., police and detectives moved across the region. On Vashon Island, officers in bulletproof vests boarded *The Typhoon* but found no one aboard. Someone saw one of Tobin's crew on a nearby boat. They held him at gunpoint. On Fox Island cops arrested four more Tobin workers as they puttered ashore in the small skiff. In downtown Seattle, not far from the Mariners baseball stadium, cops arrested another geoduck broker.

Thirty miles south in Fife, Jarmon had been watching Tobin's seafood plant and adjacent apartment since 4:20 A.M. Temperatures hovered around freezing. Light rain turned to a dust of snow and then back again. Two hours after Jarmon arrived, Volz led several officers into Tobin's warehouse. They seized .22-caliber rifles leaning against a television stand. They saw scuba tanks and dive bags stuffed with drysuits and posters of fish species tacked to the walls. Taped above a desk was a list of rules detailing how to legally ship clams. Officers found twenty-five containers of undocumented geoduck stacked on the concrete and several more clams in cold storage. The cops didn't know whether Tobin pulled the clams from polluted waters, so later that day officers would dump the geoducks in a landfill.

With two quick swings of his boot, an officer smashed in Tobin's apartment door. His oldest daughter, a teenager, stood in a hallway outside the bedroom. His six-year-old daughter ran down the stairs into her sister's arms. She'd been upstairs watching *Shrek* with her father. When the officers handcuffed Tobin and started hauling him away, he asked if he could speak briefly with Ed Volz.

Detective Pudwill retrieved Volz from the warehouse next door and brought him to Tobin, who was sitting at his kitchen table. The

two men by now knew each other well. They had worked as teammates
and opponents, had spoken civilly and otherwise. Tobin appeared to be
taking it all in, trying to think quickly and grasp the moment and his
place in it. To Volz he looked overwhelmed.

"How much trouble am I in?" Tobin asked.

Volz couldn't think of a snappy retort. "Frankly, Doug, you've had it."

Volz saw that Tobin had more to say, so he asked Pudwill to read
Tobin his rights again. The fisherman and the detective waited, staring at each other. Tobin complained that his handcuffs were too tight,
so an officer wrapped his wrists in plastic restraints. Tobin asked to
speak with his daughters, and Volz nodded. They whispered for a few
minutes, and then Tobin asked for his coat. The officers checked the
pockets and removed $2,737 in cash.

Volz took Tobin aside, and the fisherman offered to squeal on
another poacher. He said he had lots of fresh information about several new poaching operations. "I can give you the whole West Coast,"
Tobin said.

Volz stared. "Too late for that, Doug."

Volz helped Tobin find his lawyer's phone number in an address
book, and then he held the receiver while Tobin spoke to his attorney.

The wildlife cops eventually arrested more than thirty people. They
served twenty-six search warrants and seized ten thousand pages of
records at seafood companies from Seattle and Oregon to Oakland
and Los Angeles. Over the next several weeks, they visited Sea-Tac
International Airport, reviewed Tobin's air-freight bills, and gathered
documents that would prove he shipped more geoduck to Canada and
California than he bought or dug legally. Owners of a fish company in
Oregon distinctly recalled Tobin telling them with a wink how lucrative their relationship could be. He'd promised them geoducks, clams,

crab—anything they wanted he'd said he could get at great prices. In Los Angeles, a shellfish broker estimated he'd bought $50,000 to $100,000 a month of Tobin's seafood.

The detectives didn't stop there. Brian Hodgson, the original geoduck kingpin, had been arrested four months before Tobin on unrelated geoduck-poaching charges. During his trial he'd taken the stand in his own defense. Hodgson convinced the jury he'd done nothing wrong and they acquitted him. The detectives would make sure the same thing didn't happen with Tobin. They continued gathering evidence. They would make their case unbeatable.

They deployed their biologists, got Harrington working on documents, and even hired a forensic accountant. They reviewed every possible piece of paper: shipping records, sales invoices, receipts, canceled checks, wholesaler distribution records. They compared this paperwork to diving logs from Tobin's crew and GPS data from *The Typhoon*. Since the airlines kept shipping records only back to January 2000, the detectives only did accounting over the previous two years. They found that during that time, Doug Tobin and his crew had stolen nearly seventy-five thousand pounds of Dungeness crab and two hundred thousand pounds of geoduck, valued at $1.5 million. Tobin had taken nearly eight times more crab than all other recreational and commercial fishermen combined take from the southern part of Puget Sound each year—almost one-third of all the adult Dungeness crabs in the Sound's southern reaches. Given that interviews with informants suggested Tobin's poaching dated to 1997, the wildlife detectives suspected he'd made off with $3 million in contraband, probably more.

The closed geoduck beds would take thirty-nine years to recover, scientists estimated, and Tobin's shallow-water poaching near the shore jeopardized sensitive eelgrass beds that serve as aquatic nurseries for herring and endangered salmon. To protect surviving geoduck populations, wildlife managers would have to adjust overall quotas, but doing so would require a certain amount of guesswork. It would be decades before the ecological damage could be fully understood.

Sorting through the documents the detectives learned something else. They figured out why Tobin seemed so haggard that January morning in 2000 when they'd videotaped him railing about the evils of poaching. He had been out all night. Records showed that he had illegally gathered more than $13,500 worth of geoduck before coming in the next morning to lecture the camera about smuggling. The poaching tricks he'd been warning about had been his own.

Tobin waited month after month in jail while his lawyer prepared his case for trial. During that time he learned that nearly everyone involved in his operation had talked to the police. Once, Tobin used the jail telephone to call the old chain-smoking crab poacher who'd helped him find the sweet spot off the Nisqually River Delta. He listened as the old poacher admitted that he, too, had confessed. He even told Tobin the call he was making was being recorded. Tobin didn't seem to care.

Nor did he seem to hold a grudge. "Hey, how you feeling?" Tobin asked.

"Oh, I'm feeling fine, Doug, you know."

"Yeah," Tobin said. "Well, I worry about you, my friend."

"Don't worry about me," the old poacher said. "You save your own ass right now."

"Well . . . it was a good run," Tobin said. "I was sitting here grinning and I—they can never take away . . . There were some good moments with ya. I'm just glad I was part of that history, to share it with you."

"Yeah," the poacher agreed. "But I wish that the penalty wasn't going to be so damn tough."

"I know," Tobin agreed, then perked up. "I still say them was the biggest sons of bitches that ever been caught."

"Pardon?"

"Those are the biggest fuckers that have ever been caught," Tobin said. "I still stand on that."

"Who, you? Were the biggest fish that ever been caught?" his friend asked.

"The biggest *crab* that ever been caught," Tobin said.

In court, Doug Tobin crumpled into his seat at the wooden table. He was still a big man, but his pallor suggested something insubstantial. He looked drained. He had never been comfortable spending long hours indoors, and now months of living beneath fluorescent lights had taken a toll. Most of Tobin's adult life had been spent in, on, or near water, aside, of course, from his previous stints in prison. He seemed to draw energy from the sea.

Those days were gone, as were the self-effacing charm and the magnetism. The jailers had taken his rings and bracelets and his whale-tooth necklace. Instead of a flannel shirt and rubber boots he wore slippers and a jumpsuit that sagged at the neckline. Even his once-flowing silver locks seemed listless. He sported a gray goatee and had his hair pulled straight back in a high, tight ponytail. He placed his head on the table.

It didn't last long. When his head came up, his dark eyes darted about the familiar faces in the courtroom. Behind clear plastic in the gallery sat Keith Smith and Heidi Mills, such a hard-core worker that she'd returned to Tobin's employ after falling into *The Typhoon*'s crab hold and breaking a few ribs. Tobin's eyes lingered longest on his oldest daughter. She sat in the second to last row behind him, hugging a friend, her eyes damp with tears. He dipped his head slightly in her direction, an almost imperceptible hello.

Ed Volz sat to his right at the other table, beside the prosecutor. With his mustache, bolo tie, and vest, Volz looked a bit like an Old West sheriff in Tombstone, Arizona, even down to the toothpick he kept ready to twirl between his lips.

Tobin had been charged with nearly 160 crimes, including violations of wildlife law, charges of first-degree theft, criminal racketeering, and being a felon in possession of a gun. Because of his two previous convictions, he faced Washington's equivalent of a third strike. He could be sent to prison for life, and after more than thirteen months in jail, Tobin decided not to fight. He agreed to plead guilty to more than thirty crimes and face anywhere from seven to twenty years in prison. The decision would be up to the judge.

Before Tobin's plea six months earlier, his lawyer had approached Volz outside the courtroom.

"That's an awful lot of time for a bunch of fish, don't you think?" he had asked.

Volz had twirled his toothpick. "It wasn't about fish. It's always been about money."

Now it was a rainy December day in 2003. Tobin and everyone else waited for the judge. Tobin whispered to his attorney. They argued briefly, and then the attorney pushed himself from the table and walked toward Volz.

"He'd like a word with you," the attorney said, exasperated.

Volz and the prosecutor both looked up slowly.

"That so?" Volz said. "What about?"

"You should let him tell you."

The bailiff led Tobin and Volz into an empty room. The other detectives followed. Fifteen minutes later they emerged and took their seats in the courtroom. Volz and Jarmon were whispering among themselves, and Tobin looked more drained than before.

Moments later, the bailiff stepped forward: "All rise."

Pierce County Superior Court Judge John McCarthy sentenced Tobin to fourteen years in prison, the longest term in Northwest history for a wildlife smuggler. Before Tobin was taken away, the judge launched into a lecture.

Tobin's criminal operation demonstrated greed and "a great deal of sophistication," the judge said. "We often lose sight of crimes against our natural resources. These types of crimes, I think they're hard to track and probably difficult to prove, and don't have a great deal of pizzazz."

But Tobin hurt fishermen, he hurt his fellow tribal members, and he sparked untold changes to marine life in the southern Puget Sound. "The state suffers, the individuals of the state suffer, the tribes suffer, and the individual members of the tribes suffer," McCarthy said.

Offered a chance to speak, Tobin apologized to his family, in particular his mother, and thanked his attorney and the authorities who cared for him in jail. "That's it," he said, wiping away tears.

Outside the courtroom Ed Volz explained what had happened behind closed doors. Tobin had made one final pitch. He wanted to offer up one more list of poachers.

"He was mad that he got caught and there were other poachers who didn't," Volz said. "He didn't think it was fair.

"We no longer trust him. But we'll check it out."

A month after Doug Tobin began his incarceration, a bustling crowd filled the seats and lined the walls of a small conference room a few hundred yards from the waterfront in Olympia. They were gathered to witness a meeting of the three-person board that oversaw that city's port, normally a staid proceeding. The packed house included Tobin's mother and one of his daughters, but nearly everyone else was there to take Doug's name in vain.

When the port hired Tobin in 1997 to carve a totem pole for a new public plaza, the move went almost unnoticed. The port leaders who chose Tobin knew he'd been in prison, but they didn't know the details of Tobin's crimes. They viewed the sixty-six-thousand-dollar commission as another step in his rehabilitation. The agency would help a troubled, talented artist, and he would provide a fitting tribute to the region's past, commemorating the end of Puget Sound's inland marine highway. The painted, peeled piece of wood produced from the arrangement delighted everyone. It was elegant, depicting a killer whale, a seagull, and a schooner, a moon face, and an eagle in shades of red, white, black, and green. A few years after its completion, the thirty-six-foot old-growth cedar post still lay under a blue tarp in a warehouse, waiting for construction of the plaza where it would one day stand prominently against the water with the state capitol dome silhouetted in the background.

When a local newspaper story about Tobin's poaching arrest mentioned his artwork, it was an unsettling reminder to Joanne Jirovec's friends and family of Tobin's role in her 1986 death. A man who had helped kill their friend had been paid with taxpayer money to create a welcome monument for their city. It was not the civic salutation their city deserved. A few demanded his pole be soaked in gasoline and set afire.

The controversy raised questions about how or if governments procuring artwork should vet artists; about whether displaying the pole demeaned Jirovec's memory; about whether *not* displaying it disparaged the artistic and cultural heritage of the Squaxin Island Tribe, which it depicted. After all, Olympia sat on land the government had seized from its original inhabitants.

The elected leaders overseeing the port faced difficult choices. They hired an ethnologist to assess the piece, who pronounced it a "monumental" artistic achievement. The pole, in fact, had grown in value. Reducing it to embers would be like burning stacks of money. In the face of public outrage they decided to auction the piece, advertising it in art catalogs and magazines. Not a single offer approached the pole's worth.

Eventually the port impaneled a commission to figure out a course of action. A local artist and an arts advocate, a retired Washington secretary of state, the president of the Evergreen State College (where Jirovec had worked), a Jirovec family friend, and a prominent member of the Squaxin Island Tribe had planned to meet several times over the weeks to form a recommendation.

Olympia residents streamed into the meeting chambers on a January night to hear the news themselves, even though word of the panel's decision had made the local paper days earlier. A local artist stood and gave a report. The commission had voted 4-2 to urge the port to raise the pole, but only after it asked tribal leaders to bless and cleanse it in a ceremony. The pole should include a plaque with the names of all the designers and carvers who created it. Several designers had put in more time than Tobin. Any action the port took should be about the art, not the artist.

The college president and Jirovec's friend objected, arguing that the port should sell the pole for whatever it could get and use the money to commission another. "If you put that up in this community right now, that pole's going to be divisive until after my grandkids are gone," Jirovec's friend Jim Snell said. "And I don't know if you want to live with that legacy." Snell said raising the pole would lend Tobin distinction, "and he does not deserve honor in this community." He held up two fingers a quarter inch apart. "Not even *that* much."

Tobin's daughter reminded the port that it had hired her dad because he was a significant artist who had artwork displayed in galleries and museums. One man reminded those present that even the Republican White House had just encouraged prisoners to reenter society. "Those of us who heard President Bush in his second State of the Union address mention second chances for people reentering society after having served their time might understand the [Port of Olympia's] frame of mind when they signed a contract with a known felon," said sailor Paul Deranleau. "Who was to know that Mr. Tobin would return to bad actions? Art often transcends the character or behavior of the artist. Indeed we would have little in the field of art, music, or dance to enrich our lives if only the righteous were represented."

Jim Peters, a member of the Squaxin Island Tribe, didn't dispute that Tobin's actions were inexcusable. But he reminded everyone in the room that Native Americans every day compartmentalized their own history. It was the only way they could live among so many monuments to settlers who had perpetrated violence against their ancestors. "There are a lot of things that have happened that I am told that I need to forget about on a regular basis," he said, "and I do, because I have to."

Still, Snell had set the tone for the evening. One by one, community members stated their objections to the commission's recommendation. Some read letters from Jirovec's children and siblings. A family friend read a newspaper editorial. Another suggested the pole had split the city already. One woman called it a "slap in the face" to law-abiding citizens and suggested it be dumped in the Sound and allowed to float away. A man who knew Tobin when he poached said, "Putting up a plaque that recognizes Tobin in any way only gives him the notoriety he seeks."

The port overruled its volunteer panel and decided to get rid of the pole, even though one port commissioner said, "It's one of the most beautiful things I've ever seen." In the end, it was no longer about the art, another port official said. "It's about Doug Tobin." Within six months the pole was gone.

But neither the pole nor Tobin was forgotten. Hearing the pole would go to auction again, the director of the University of Washington's Burke Museum of Natural History and Culture in Seattle nudged a wealthy collector of Indian art to pay thirty thousand dollars for the pole and then donate it to the Burke. This donation was not well received by Jirovec's friends. "Can you imagine the City of Modesto putting up a monument made by Laci Peterson's killer?" one asked. The University of Washington had been Jirovec's alma mater.

Eight months later, authorities hauled Tobin back into court. A diver had belatedly told detectives that while Tobin was poaching he'd helped firebomb a rival fisherman's boat. The new accusation emerged from Slim, the six-foot-six diver from Tobin's crew. Slim had been quite forthcoming after his arrest. He'd filled the detectives in about the crab and clam poaching, and about the identity of other crew members. The detectives saw Slim as a decent kid seduced by the promise of riches. They encouraged the prosecutor to let him go home to Port Angeles while he awaited his own court case. A shot at a light sentence hinged on his cooperation. In return the detectives demanded he keep in touch with the prosecutor. Slim did so devotedly for a while.

Then one day, Jarmon heard from the prosecutor that Slim had not phoned in. Jarmon was annoyed that he'd been played and he wanted an explanation. He and Volz confronted Slim at his home and called him a liar and told him the deal was off. Slim looked stunned and said that he'd left the prosecutor several messages. The detectives asked what other lies he'd told, what else he knew. If he had more to say, now was the time.

He told a jarringly familiar story. Sometime after midnight on September 4, 2001, Tobin and his brother had been looking for a place to steal geoducks. They wound up not far from the Purdy Spit, where the brothers first met their diving instructor. They recognized the wooden *Laurie Ann* docked offshore. *The Typhoon* pulled up, and Slim said Tobin ordered him to hold the boats together while he and John stepped aboard the ninety-year-old ship. The brothers slipped into the engine room, cut open a fuel line, and soaked and lit on fire a swatch of carpeting. The brothers jumped back aboard *The Typhoon* and pulled away as smoke billowed from the *Laurie Ann*. Doug then leaned in to Slim and said, "This never happened."

The detectives shot back toward Seattle, thinking about Dave Ferguson. Another boat had been blown up over clams—at least that's how it looked. They met again with the prosecutor, who sheepishly confessed that he might have misplaced Slim's messages. The detectives held their tongues.

Back in 2001, the fire had not even been considered suspicious. A fireboat called at dawn that late summer morning had found the *Laurie Ann* engulfed in flames. The orange licks reached too high to douse from the water. Firefighters towed the burning husk to shore and boarded it with ladders, but they were too late. Investigators presumed that the fire had started in a malfunctioning electrical panel. A deputy fire marshal photographed the boat's interior, planning to give it a harder look later, but slipped in greasy water and ruined his camera.

The *Laurie Ann*'s owner had infuriated lots of people, particularly tribal divers, and was in a legal skirmish with the government of a local county over millions of dollars in geoducks. He'd been experimenting with planting geoducks much like shellfish companies farmed oysters and had leased forty-seven acres of public tidelands just off a county park near Purdy Spit. Since the land was underwater and not suited for building, the county charged just twenty-five hundred dollars. But in readying his farm for planting baby clams, the *Laurie Ann*'s owner had set about removing obstacles—namely an entire colony of wild geoducks. For the bargain price of twenty-five hundred dollars, the *Laurie Ann*'s owner had dug up and sold $2 million worth of clams. Angry neighbors complained that he was openly poaching, and it seemed plausible to the detectives that Tobin or his brother might hold a grudge, too.

The Tobin brothers went to trial in 2005. One of Doug's crew members confirmed Slim's story, but another maintained two other men set the fire. The fire marshal had initially ruled the fire an accident but later decided it could have been arson. Still, his photographs were gone, and the boat's remains had since been scrapped, leaving no hard evidence for anyone to examine.

John Tobin had a friend ready to testify that he had been fishing the day of the fire. Doug Tobin claimed he was at a car show in Indiana. Defense attorneys sliced the case to ribbons. Both were acquitted.

It would not be the last battle over a boat.

Detective Volz wanted *The Typhoon*. The court had ordered Tobin to pay more than a million dollars to repair the damage he'd done to southern Puget Sound. But Tobin had either hid his earnings or spent them. Between his jewelry, his daughters' private schooling, and his five-thousand-dollar-a-month restaurant bill, Tobin had expensive tastes. He'd put more than a hundred thousand dollars into a souped-up Nova, but that had been stripped by friends who took the parts. Though the cops heard rumors of offshore accounts and of real estate holdings in someone else's name, no one had been able to find the money.

But *The Typhoon* was worth something. After detectives seized it, the forty-two-foot boat sat on blocks in a state impound lot off a southern Puget Sound beach. It had been Tobin's most precious possession and perhaps the most valuable one; at one time it was worth at least four hundred thousand dollars. But the state could do nothing with it. The Canadian company run by Jeff Albulet and Julian Ng held liens on the boat. It could not be sold to pay Tobin's restitution unless the Canadians released their interest, which they refused to do, because they believed that Tobin owed them money.

But over just the last two years of his operation, cops found evidence that Tobin had sold geoducks to the Canadians at least 150 times. He'd sold them 135,000 pounds of clams. He also had sold more to other customers at their suggestion. The Canadians had fronted Tobin thousands of dollars to keep cheap shellfish coming over the border. Yet they insisted they'd never known Tobin was acting illegally. "The man, as an individual, seemed like a nice, funny person," Albulet said shortly after Tobin's conviction. "It's just that he knew how to con people. Unfortunately, he got me for a lot of money."

Neither the detectives nor Washington State's attorney general believed Albulet. The detectives asked around. Adrian Lugo had found Albulet and Ng fierce negotiators and well versed in the legal intricacies of shellfish brokering. One of Tobin's daughters and several other people told the cops that Albulet had been on Tobin's boat several times at night while Tobin poached. Heidi Mills remembered people at the plant hearing from both Canadians that they wanted to know if the "mystery boat" from the Tacoma newspaper was *The Typhoon*. Albulet knew, Heidi Mills told the detectives. "He's no dummy." Tobin's brother said he'd attended a meeting when Doug told the Canadians he was getting their product illegally. Mills, Lugo, Doug's daughter, and John Tobin all signed affidavits detailing the Canadians' role in trafficking.

But prosecuting a Canadian company was difficult. The state could file a lawsuit and try to force the Canadians to relinquish their interest in *The Typhoon*, using the civil equivalent of the racketeering and fraud laws that had forced Brian Hodgson sixteen years earlier to shut down Washington King Clam. There was no guarantee how it would turn out, but the attorney general's office would give it a shot. There was just one more person to interview.

On a damp summer afternoon years after Tobin was sent to prison, Ed Volz pulled his pickup onto a side street in Tacoma and walked a few blocks to the county jail. The floors shined in the new towering facility, even though the county was too short on cash to hire corrections officers. It had been forced to grant early release to nearly three hundred felons since January.

Volz was joined by two assistant attorneys general. They had convinced the Department of Corrections to let them interview Tobin inside the jail. It had not come without a fight. Before agreeing to talk, Tobin's lawyer wanted guarantees that his client would face no new charges. He tried to ban the detective from the meeting. Volz and the AGs were so suspicious of Tobin that they didn't expect anything useful to come of the conversation. They just wanted to assure themselves that they had pursued every angle. They didn't want any surprises, so they told Tobin's attorneys they would talk only with no strings attached.

The attorneys assumed Tobin agreed to the meeting because he felt burned by the Canadians. Volz thought Tobin *liked* talking to the authorities. He was a player. He wore lots of jewelry, drove fast cars, and carried thick rolls of bills in his breast pocket. And he liked matching wits with cops. Volz figured that was the juice for Tobin: going mano a mano with opponents. But who did Tobin consider his opponent now: Volz or the Canadians?

Volz and the attorneys headed into a small meeting room with glass windows and a view of the hall. Tobin looked nothing like the attorneys expected. Instead of magnetism, generosity, salesmanship, and charm, they saw a pale, broken lump in jail slippers and a jump-suit. Tobin looked so gray and sickly they feared he'd been drugged. At one point he stepped outside to receive medication from a nurse.

Volz's team started the meeting with a speech. The state wanted every scrap of information it could get on the Canadians. But given Tobin's track record, everything he said would be worthless without corroboration. Unless Tobin could point them to independent evidence to support any accusation, nothing he said would make a difference.

They emerged from the meeting at 4 P.M., tired, but upbeat. Tobin claimed Albulet had been on the boat with him poaching at night more than twenty-five times. But many of his other recollections did not fit with the evidence. On the few occasions when the attorneys pointed that out, Tobin said things like, "Hmm, let's think about this. If we all put our heads together, we can find an explanation." Talking to Tobin hadn't helped them much. But he hadn't said anything to torpedo their case.

The matter settled sixteen months later. Volz and the state decided that even if they went to court and won big, the whole thing could back-fire. Albulet and Ng were foreign defendants, and their company was incorporated on Canadian soil. Any judgment in a Washington court would still have to be enforced by Canadian courts. The two men had threatened that if the state won, they would declare bankruptcy and no one would see a dime. The attorneys were unfamiliar with Canadian bankruptcy laws and did not consider it an empty threat.

So they settled the case for less than the attorneys and Volz thought it was worth. But the Canadians were banned from ever again participating in Washington's shellfish industry. The state could keep and sell *The Typhoon,* and the Canadians would pay an additional $112,500 in restitution. On the day the state received the check, one of the attorneys made an extra photocopy so he could hang it on his office wall. Along with the Brian Hodgson case in the late 1980s, the two biggest wildlife poaching rings in Pacific Northwest history had been about geoduck clams. Now both had ended with poachers behind bars and big checks cut by the businessmen who profited.

It wasn't a million bucks or even full restitution. It would provide barely any help for Puget Sound. But it sent a message. At least that was the hope.

Epilogue

"It was like a Walt Disney movie that turned into Stephen King," Doug Tobin said. He looked around, waggling his nose, as if catching a whiff of something sour. Tobin was talking about the years he had spent harvesting geoducks, but he might also have been referring to far more.

He sat at a wood table, an empty water bottle before him. Around him families chattered the way they do at holiday gatherings or Sunday brunches. Tobin spoke of himself as if he were somehow separate from those around him. I understood the impulse. This wasn't Thanksgiving, and no one was celebrating. It was just another Sunday, visiting day in prison.

I had driven up from Olympia on a wet slate-gray morning. I motored past hundreds of salmon fishermen wading elbow to elbow in the rain, so many in one estuary that I heard the spring-loaded snap of a door to a plastic Porta Potty clang shut. Authorities had dumped it there to discourage peeing in the woods. I stopped for gas at the Kamilche Trading Post, a convenience store and gas station run by the

Squaxin Island Tribe. Another few miles and the woods gave way to grass fields and strip malls along the outskirts of Shelton, a quiet mill town. A sign pointed the way to the four-hundred-acre Washington Corrections Center compound.

Except for the razor wire and guard tower, the prison looked from the outside like a forgotten Cuban piano bar, a single-story square of white-splashed concrete with brick latticework and aqua trim. Inside, I stuffed my belt, wallet, and keys into an aluminum box in a bank of lockers and stepped through a metal detector. Past two gates I ended up in a sterile room the size of a hospital cafeteria. A guard stamped my wrist like a club bouncer, leaving a pattern visible only under the black light on her desk. My exit pass. She looked over a seating chart, sent me to a corner table, and told me to wait.

About two dozen square tables filled the bulk of the room, each with four cheap stackable chairs. One wall held a bank of six vending machines. A corner housed a children's play area lined with mats and plastic toys, the walls painted to look like a desert island surrounded by warm blue waters. Families played Scrabble, poker, and board games like Clue. Some inmates wore jeans, others shuffled about in jumpsuits with wcc in block letters on the back. The prisoners all sat facing in the same direction. At the far end of the room family members could stand before a wall designed to look like a waterfall flowing through a rain forest while a guard snapped portraits with the felon. I tried not to look anyone in the eye.

Prisoners entered and exited through a door that looked like it belonged on a submarine. After about ten minutes, Tobin strolled through. He wore jeans and a zip-up khaki sweater over a white T-shirt. His hair was still long, but he looked forty pounds lighter than he had years earlier. Seeing me, he ducked playfully behind a pillar, waited a few beats, then snuck a peak around it with just one eye, like a grand-father teasing a toddler. After a few seconds he smiled broadly and sauntered my way. I lost my hand in one of his huge leathery mitts, and before I could speak he deadpanned, "Let's go do some shopping at the mall."

Offenders could not touch the vending machines; visitors used special prison-issue debit cards to treat them to Cokes and Snickers. I fiddled with the turnstiles while Tobin frowned at his options from behind a line taped on the floor. He pointed to a ham sandwich—"Just give me whatever that is"—which I bought and stuck in a microwave. I couldn't help but notice how powerful he looked. Now in his midfifties, he appeared lean and fit as a boxer. On his jeans he wore an ID card with a mug shot of an earlier, bloated Tobin from the start of his incarceration. The man before me looked twenty years younger.

A phone at the guard station rang, and Tobin joked, "I think that's gonna be for me." The female guard threw him a pleasant, mischievous smirk and grabbed the phone with a singsong: "Tobin's Answering Service." They both laughed, and she turned her attention to the call. Tobin looked at me with a nod and a wink.

"They like me around here," he said.

I had interviewed Tobin shortly after his arrest, just hours before he pleaded guilty. He had bemoaned his predicament and hinted at what he'd seen. "This much I can tell you with one hundred thousand percent accuracy," he'd said in a near whisper. "I can only think of one or two people who are on the same level I'm at in this geoduck industry. I've seen it all— everything from prostitution to dope to contract killings. Everything you can imagine. Everything for a goddamn movie." After I'd revisited the events of the previous decade, I wrote asking to see him again. I wanted to hear how he thought he had wound up behind bars. He wrote from prison encouraging me to come. "I do believe you could use some insight into the geoduck world—the good, the bad, the ugly," he wrote. "I can be reached at my office 24-7 . . . so let's dance." In an earlier note he'd referred to himself as "The Geoduck Gotti." This one he signed "Elvis."

We returned to our seats and chatted while he ate. He opened up about his childhood and his years in rain-soaked Alaskan coastal villages. He spoke respectfully of his father even while describing how hard the man had pushed him. His family had moved a lot, and so had Tobin as an adult. "I was kind of a nomad, I guess," he said. He attributed his love of muscle cars to being born in the 1950s and to having fished and logged at a young age. "I never had much of a childhood," he said. "I didn't really get to be a kid."

Doug Tobin describes the geoduck industry during an interview in jail.

He boomeranged between topics. He recalled wilder episodes of the tribal fishing wars, such as the moment he watched a woman level a gun at a marine officer. Then he returned to discussing fast cars and his youth. He laughed about his poaching exploits as a young man. He launched into long digressions, including one about cutting open an ice-cream truck with a blowtorch so he could fill it with salmon to sell at Seattle's fishing port. They were good stories, funny and well told, if rarely on point. He made frequent and direct eye contact and nodded often, as if to acknowledge his wandering and the fact that I was waiting. "I've got to tell a story to tell *the* story," he would say and hold up a hand, as if seeking patience.

He agreed to be an informant because he saw fishermen with "greed twinkling right out of their asses." He insisted he had started geoducking on the level, "but by then you were so choked with all the corruption, all the madness, the conning." For a moment it seemed as if he would explain how he had crossed the line, but instead he said, "So I reached out to try and find some way to help clean the industry up."

Tobin spoke fondly of federal agents, describing them as a surrogate family. When Severtson outlined plans to zero in on smugglers, Tobin said, he'd reminded the agent that he already knew how best to play it. "I told him, 'I know these people; your way isn't going to work.'" Tobin said he recommended the cops put him out front, letting them "use my past as a shady individual to get into doors they could never get into." He wistfully recalled Severtson pretending to be a tourist while secretly videotaping a poacher. He recounted Agent Dali Borden posing as his half sister and crashing at his home when others arrested Nick DeCourville in Vegas. Tobin said he felt good working with the cops. It gave him a nice feeling giving something back. He liked doing good without asking for a return. "Money means nothing to me," he said.

I suggested I found it difficult to know when he was telling the truth. Once, he nodded slightly, as if to say he understood. Another time he exploded. "I'm not blowing smoke up your ass. These are facts!" I began to sense my own foolishness. I had wanted to hear how someone of such talent kept getting himself into such messes. He knew the geoduck industry well enough to have made it rain money legally. Adrian Lugo had all but promised him wood-carving contracts that would have paid more per piece than most people made in a year. Everyone who knew Tobin spoke about his hunger to succeed; he hated being less than exceptional at anything. But he remained unwilling to engage in self-reflection, apparently reserving the worst deceptions for himself. Perhaps his half-truths had just become part of him.

In a strange way he embodied the clam that had defined his last decade: large, obscene, and ugly from one angle; comic and charming and clearly part of the fabric of the Pacific Northwest from another. That charm worked on federal agents, a highly successful entrepreneur, and government officers at the Port of Olympia, until in each instance a darker side emerged. During our many talks, I found Tobin a mix of good-natured cheer and ferocity, a firecracker with a short fuse stuffed inside a pastry. In the middle of a story his affability would evaporate, and he would rage about fellow fishermen, DeCourville, and, most often, state detectives. He referred to the detectives by his own special nicknames for them—the jovial Harrington was "Hee-haw Hound Dog"; the toothpick-twirling Volz was "Fast Eddie" or "Hollywood Ed." Mostly he complained that they just weren't that good. "They're somewhere between whale shit and the bottom of the ocean," he said of the cops. "If I was going to poach, they'd never know it."

Tobin admitted the detectives had caught him. But he insisted he'd gathered only about six thousand pounds of geoduck and three thousand pounds of crab, little more than a few nights' labor and a mere fraction (less than 3 percent) of what the judge ruled he owed restitution for taking. Tobin's calculation didn't jibe with the confessions, the months of surveillance, the diving logs, or the eleven thousand pages of evidence against him. It wasn't even clear it was really what Tobin thought. Not two years earlier he'd told Volz and the attorneys that one of the Canadians had ridden aboard *The Typhoon* at least twenty-five times while he was poaching. With me, Tobin insisted he had been railroaded.

Then, almost as quickly as he would work himself up, Tobin would be polite and funny again, apologizing for his outburst. "I've got a good heart," he said after one long silence. "Even my worst enemy knows that's true."

━

Detective Volz's partners were mostly gone. After Tobin's arrest, Detective Bill Jarmon got promoted to deputy chief for state wildlife enforcement, where he oversaw detectives and all the marine patrol officers and game wardens. From his office, he could have walked over and visited with the governor, but he wouldn't do so unless work required it. The new pay was nice, but mostly the job ate at him— the politics, dealing with lawmakers, dealing with other agencies, worrying about budgets. He missed going after bad guys. His father had been a cop and so was Jarmon's son, and now suddenly Jarmon himself was an administrator. But after three decades in law enforcement, he knew he could stomach it for a little while, which he did. Then he retired and started spending part of the Northwest's long, wet winter driving along sunny Arizona highways in the family RV.

Kevin Harrington retired, too. Ed Volz and Charlie Pudwill helped plan his going-away bash. They put on a PowerPoint slide show of Harrington's life. They started with photographs of an infant in a crib, a cigarette dangling from his lips. The next shot was of a puppy, a cigarette poking from between its paws. Harrington wrote a song for the occasion:

> *When I go to Olympia, there's danger everywhere*
> *I forget to wear my gun, they think that I don't care*
> *But I am just forgetful, so I quit coming 'round*
> *Now that I'm retiring, please quit patting me down*
> *Retirement, retirement*
> *Sounds like lots of fun*
> *But how do I tell Bill Jarmon*
> *That I can't find my gun.*

After retiring, Harrington moved back to Michigan, where he kept writing songs. He also spent time fishing midwestern lakes. On trips to Seattle to visit his son and college-age daughter, he still groused about the traffic.

Rich Severtson and his wife moved to a place in the mountains near the Canadian border in eastern Washington, an area thick with bear and deer. A river flowed through their backyard. Severtson toyed with writing a book about his career, or maybe just about Doug Tobin. He kept copies of videotapes from his undercover work and copies of the taped phone calls between Tobin and DeCourville. He was just waiting for time to pass so he could write up his experiences without triggering the wrath of the National Marine Fisheries Service.

Then, almost overnight, Severtson got sick. He developed fast-moving and painful pancreatic cancer. In early 2004, less than a dozen weeks after his diagnosis, Severtson died in his sleep. Paul Watson and his Sea Shepherd Conservation Society issued a press release mourning his passing and highlighting his accomplishments. "Rich was a man with a great love for nature and was a fierce protector of marine wildlife," the group wrote. "He will very much be missed." The activists had never done such a thing for a federal cop, and they have not done it since.

Detective Ed Volz earned a promotion, too. He was bumped to lieutenant and then to captain and put in charge of a crew of mostly new detectives. He oversaw long-term investigations and undercover operations. He still saw Jarmon from time to time, and he missed butting heads with Harrington. In fact, he needed a guy who was comfortable around paperwork. His new charges were more action-oriented.

The last time I visited Volz he still worked from the same corner office in the same whitewashed warehouse, though the nearby Boston Market was now an Azteca. He had his feet on his desk while he watched a blip appear intermittently on a map on his laptop. He was following a signal from a transmitter attached to an illegal wildlife buyer's car as it tooled around Puget Sound.

It had been a busy spring. His colleague, shellfish biologist Don Rothaus, had been working with scientists from the University of Washington in a desperate attempt to save Puget Sound's abalone population. Scientists had been breeding the creatures in a laboratory. Rothaus planned to take the juvenile shellfish out that summer and transplant them into waters near the San Juan Islands. Researchers in British Columbia were doing similar work to restore abalone, but nobody knew how the projects would fare long term. Volz worked a dozen feet from Rothaus's cubicle and had been following the effort closely.

Volz's detectives had also broken open a black bear gallbladder-smuggling ring in the mountains not far from Severtson's former home in eastern Washington. A poacher had been selling dozens—possibly hundreds—of illegal gallbladders overseas. The overseas buyer, a wealthy big-game hunter, had previously been caught sneaking hummingbirds out of Ohio. Now he was in the States, perhaps to set up a meeting. Volz, from his desk, was watching blips on his screen, tracking the buyer's every movement.

Volz had been on the job for more than three decades. Sometimes, when he thought too much, he feared he'd made a mistake. He feared he had lied to himself that he made a difference. He wasn't a cop at heart, but a biologist. He'd watched over the years as salmon and rockfish and other marine creatures continued to decline in Puget Sound. At least now there was talk of trying to clean it up. He wondered sometimes what he'd accomplished. Then the phone would ring and jolt him back to the work he'd chosen for good or ill.

He had taken a particularly strange call in 2006 from Mexican government officials who wanted his help. Fishermen working the golden sands along Mexico's Baja peninsula had discovered a few fields of naturally occurring geoducks. One field blossomed along the Sea of Cortez. The other showed up along the Pacific Coast, among the sugared sands and pearly lagoons of Magdalena Bay, where a series of sandy barrier islands provided gray-whale breeding grounds.

The geoducks they had discovered were beautiful with scrubbed white shells, though the necks seemed lifeless and a bit flaccid, perhaps because of the warmer waters. The bay certainly produced incredible shellfish—abalone, blue crab, lobsters, shrimp, and scallops. But no one had expected to find geoducks so far south, especially with no other population source nearby. The leaders of southern Baja's government recognized an opportunity and had wanted to establish a small geoduck industry, perhaps to supply lower-quality geoducks. But some authorities had handed out permits like cigarettes. Rumors spread of bribed Mexican officials, and poaching in Magdalena was already widespread.

That spring, a group of Mexican cops and fisheries officials traveled north to meet with Volz and the agencies that managed geoduck fishing. They wanted to find ways to fish geoducks sustainably. To his astonishment, Volz learned that some Mexican fishermen were backed by familiar troublemakers from Washington and were smuggling the geoducks into the United States and then shipping them to Asia. Volz volunteered to send the Mexicans home with court documents so they could keep tabs on visiting Americans.

"It never ends," Volz said from his office, shaking his head. "It just changes." I had already asked Volz if he still saw much clam poaching, and he said that he had cases working that very moment. But now his attention was divided between me and the screen. We both sat for a moment in silence and watched his bear-parts buyer move through the city, the blips popping up as the man moved down the road.

Notes on Sources

My interest in wildlife crime grew from a small story I read in the *Seattle Times* about the arrest of five poachers who had stolen more than $3 million in geoducks. I was the paper's environment reporter, and I was floored. Who poaches clams and who hunts clam poachers? Pretty quickly I learned this wasn't the first arrest of clam smugglers. A case in the 1980s was, at the time, the biggest white-collar fraud investigation in Washington history. Another geoduck smuggler had hired a hit man to muscle a rival. One poacher had worked as a federal undercover geoduck-poaching informant. I was hooked.

Over the next few years I wrote a few stories about geoducks and about poaching, and I found myself noticing other strange plant and animal crimes. There was the theft of thousands of dollars in mosses from Mount St. Helens National Monument. Then there were the millions of dollars in grasses and shrubs that were taken illegally from national forests. Those crimes included shootings, armed robberies, and men threatening one another with shovels and machetes. One rural county employed a deputy sheriff who policed poachers full-time. He sometimes made a hundred arrests a year. I wrote about the secretive $100-million-a-year trade in wild furs and a Justice Department investigation into charges of price fixing and collusion among buyers of otter pelts. One thing linked most of these cases: The stolen plants and animals were usually bound for markets overseas.

Individually, the crimes were strange and interesting and in some cases—
though not all—added up to real ecological trouble. I was glimpsing an interesting
pattern. I called experts with TRAFFIC—the worldwide nonprofit organization
dedicated to cracking down on wildlife smuggling—and the U.S. Fish and Wildlife Service, who agreed that international wildlife smuggling was in transition.
The types of creatures being stolen and the methods used by poachers were
expanding. The number of goods illegally taken from the United States—while
still much smaller than the number of goods shipped into the country—was on
the rise. You could see it in the clams leaving Puget Sound for far shores or the
turtles plucked from Pennsylvania woodlands bound for Asia. China's influence
was increasing daily.

I was interested in looking into that world, but not from some exotic locale.
Killing elephants illegally for their ivory was horrible, but it was unfortunately no
longer surprising. More surprising was the scale of investigations taking place
right here, in cities like New York, Los Angeles, San Francisco, and Seattle. I
thought I could draw readers in with a fresh look at a story they thought they
already knew, especially if I could tell it through a character most had probably
never seen, one of the world's least charismatic creatures—the geoduck.

The foundation of this book is more than twenty-five thousand pages of public
records, mostly from state and federal criminal investigations, some of which
were unsealed by a federal judge at my request. Those records include: videotaped surveillance, audio recordings of interviews with suspects, recordings of
undercover telephone calls, transcripts of interviews and undercover conversations, computer disks of undercover video conferencing chats, depositions, declarations, court transcripts, and the painstakingly detailed daily narrative reports
filed by dozens of state and federal agents.

I also interviewed more than one hundred people—some many times—
including nearly every major character in this book and many whose names
appear nowhere in it. A partial list can be found on pages 269–272. Many went
far out of their way to help, none more so than Detective Ed Volz. He put up with
dozens and dozens of hours of interviews over the course of several years. Never
once did he ask how I planned to use what he told me. He and Detective Bill
Jarmon drove me around to each of the spots where they conducted surveillance
on *The Typhoon*. Detective Kevin Harrington took time away to visit with me

while vacationing in Seattle with his son and daughter. In San Francisco, Special Agent Roy Torres drove me around San Leandro and showed me Reverend Kevin Thompson's home, his church, and the True World Foods distribution center where Thompson's fishing crew stored illegally caught sharks. Before I visited special agents Andy Cohen in Massachusetts and Al Samuels in South Carolina, both went through their own notebooks many times to track down tidbits, jog memories, or dig out contemporaneous observations of ten-year-old events.

Three biologists at the National Marine Fisheries Service forensics laboratory in South Carolina gave my family and a friend a three-hour tour of marine pathology none of us will soon forget. Special Agent Ed Newcomer put up with far too many prying phone calls from me asking about how his family felt about his work undercover. Geoduck expert Brent Vadopalas never refused to meet for lunch, or explain for the tenth time an obscure point of shellfish biology, or read a paragraph whose accuracy concerned me. Special Agent Sam Jojola told me about his stash of secret identities and made introductions on my behalf to several legendary undercover investigators. Special Agent Lisa Nichols drove me to the California-Mexico border at San Ysidro and helped me understand how complicated it can be to catch someone sneaking across with wildlife. Paul Watson, with the Sea Shepherd Conservation Society, gave me his satellite phone number and took my call while on break from chasing whaling ships in Antarctica. Bob Donegan, president and chief executive officer of Ivar's, showed off Ivar Haglund's collection of paraphernalia and tracked down sixty-year-old menus and videotapes of old commercials. Carl Sheats gave me all of his father's papers, including letters the elder Sheats wrote in the 1960s describing his initial discovery of geoducks in the deep. Claude Tchao sat through several hours of telephone interviews after a freak snowstorm shut down transportation between Seattle and Vancouver on the very week I had planned to visit him. Robin Wright, curator of Native American Art at the University of Washington's Burke Museum of Natural History and Culture, drove me to a warehouse to show me the welcome pole carved by Doug Tobin, and enthusiastically filled me in about its style and quality.

Doug Tobin, too, never refused a visit, agreeing to meet with me many times and putting up with me even as I asked the same questions over and again. Special Agent Richard Severtson and his wife invited my wife and me to stay with them for the weekend. In his living room, he played tapes of telephone calls he'd recorded between Tobin and Nichols DeCourville. His wife made us dinner, and he urged us to fish the stream that cut through his yard. He never once mentioned he was sick.

I also relied on countless government reports, from congressional hearing minutes to a decade's worth of annual reports by the law-enforcement branch of the U.S. Fish and Wildlife Service to dozens of competing studies about overfishing around the world.

Details about the case built against Brian Hodgson were drawn almost exclusively from several thousand pages of a twenty-year-old police investigative file that attorney Marilyn Brenneman retrieved for me from storage—more than forty legal boxes' worth. The chapter about shellfish and the U.S. Exploring Expedition was almost entirely reported at the Ernst Mayr Library in the Museum of Comparative Zoology at Harvard University. Details of how geoduck and other contraband move illegally into and within China came from U.S. government documents and Chinese news reports. Ting-I Tsai, a freelance journalist based in East Asia, tracked down and translated news and documents written in Chinese.

Throughout the book, thoughts and actions attributed to characters are from interviews or public records or correspond to what the person said he or she was thinking at that moment. The same holds true for gestures; when I wrote that Volz "pancaked both hands against the dash" it's because that's how he recounted the event. Extensive quotes or exchanges of dialogues are almost entirely drawn from video or audio recordings, official transcripts of interviews, or undercover recordings. Some were documented in notes and letters or in police reports, court testimony, or depositions. In a few cases, short quotes are recalled by participants as the precise language used at that time. I actually witnessed a handful of the scenes near the book's end.

〜

PROLOGUE: THE HUNT

1   *The boat didn't look like much:* I walked with the detectives the route they traveled on November 12–13, 2001. They carried the narrative reports they'd filed just days after their reconnaissance. We scrambled over the blackberry brambles, walked the eroding beach, and Volz showed me the precise spot where he had lain in the fallen leaves, looking down on Wyckoff Shoal. Descriptions of the boat came from surveillance videotapes and photographs. I also toured and photographed the boat inside a Washington State Patrol impound lot. Information about how it was maintained and operated, and how much detectives knew on November 13, came from my interviews and transcripts of taped interviews the detectives had with confidential informants Keith Smith, Heidi Mills, and Mark Purdue. I possess transcripts of every witness interview the detectives conducted between January 2000 and December 2005.

3   *Coarse-grained sand coated the bottom:* C. Lynn Goodwin, interview with author; Alex Bradbury, interview with author.

4   *An informant had described crew members:* Keith Smith, transcript of taped interview with Detective Bill Jarmon, August 2001.

4   *Four times in two weeks:* Reports of surveillance by Ed Volz, Bill Jarmon, and Charlie Pudwill, on October 25, October 30, November 6, and November 8, 2001.

5   *couriers boxed geoducks with gel packs:* Claude Tchao, interview with author; Casey Bakker, interview with author.

5   *traded for narcotics:* Special Agent Richard Severtson, interview with author.

6   *Geologists call such cliffs:* My understanding of the term "feeder bluff" grew out of years of reports on shoreline ecology prepared for the Puget Sound Nearshore Ecosystem Restoration Project, a joint rehabilitation effort by the U.S. Army Corps of Engineers and the government of Washington State.

CHAPTER ONE: SNITCHES

9   *Ed Volz slumped in a car:* I watched Ferguson collect the abalone from his Jeep on one of several video clips shot by wildlife detectives during their surveillance on June 17, 1994. Most of the video was shot outside the Chinese restaurant. Details of the conversation that took place inside came from two sources: notes typed up days later by Detective Bill Jarmon, who watched the discussion through a mirror behind the bar, and transcripts of the recorded interview Ferguson gave detectives that night. Every little detail, from the refilling of their drinks, to the topics the men discussed, to how the abalone buyer was dressed, came from those sources. The quote—"You will learn and I will learn and we will learn together"—is recorded as a direct quote in Jarmon's notes.

11   *The sheltered marine waters:* Geoducks can be found from Southern California to southeast Alaska. Even as far north as Oregon or the outer coast of Washington, the density of clams is not enough to support any fishing industry.

12   *Elephant tusks, wild furs:* All of the smuggling cases mentioned in this section are real examples. The bulk of them came from four places: *U.S. Wildlife Trade: Overview for 1997–2003,* Office of Law Enforcement, Intelligence Unit, U.S. Fish and Wildlife Service; the annual reports, Office of Law Enforcement, U.S. Fish and Wildlife Service, 1995–2006; *U.S. Illegal Wildlife Trade, LEMIS Data Analysis and Risk Assessment,*

U.S. Fish and Wildlife Service, November 2005; and interviews with more than a dozen federal agents.

12 *Baboon noses:* A shipment of two thousand rotting noses bound for the United States was seized in Amsterdam in 2003. See: *The TRAFFIC Report* 3, no. 1 (March 2004): 11.

12 *cycads:* Lauren Kessler, "The Cult of the Cycads," *New York Times Magazine,* August 28, 2005. Special Agent Lisa Nichols showed me photographs of birds stuffed in pickup consoles, taped to smugglers' bodies, packed into badminton birdie tubes, and shoved in curling irons and helped me find the indictment of a thirty-three-year-old Long Beach man for hiding iguanas in his prosthetic leg.

12 *eels:* Todd Shields, "Eel Poachers Try to Slither Past Police; Tiny Fish, a Delicacy in Asia, Protected Here," *Washington Post,* March 30, 1997; and Hannah Hoag, "Eels on Slippery Slope," *Toronto Globe and Mail,* March 31, 2007.

12 *seismic shifts in the world economy:* My grasp of the economic forces that fuel the global underground was aided by interviews with Crawford Allan of TRAFFIC; Craig Hoover with the U.S. Fish and Wildlife Service; and Nancy Foley, chief of law enforcement for the California Department of Fish and Game and one who has studied the trends extensively. Their intuitions were confirmed by several studies, including "Shadow economies: Size, causes, and consequences," *Journal of Economic Literature* 38, no. 1 (March, 2000): 77–114; and, by the same authors, "Hiding in the Shadows: The Growth of the Underground Economy," Economic Issues Series, no. 30, International Monetary Fund, March 2002. Two books were also invaluable: Moises Naim, *Illicit: How Smugglers, Traffickers and Copycats Are Hijacking the Global Economy* (New York: Doubleday, 2005); and Eric Schlosser, *Reefer Madness: Sex, Drugs, and Cheap Labor in the American Black Market* (New York: Houghton Mifflin, 2003).

13 *illegal trade accounted for 10 percent of the world economy:* Friedrich Schneider and Bruno S. Frey, "Informal and Underground Economy," in *International Encyclopedia of Social and Behavioral Science,* vol. 12 (Amsterdam: Elsevier Science Publishing Company, 2001).

13 *one of the largest black markets in the world:* State Department, Bureau of Oceans and International Environmental and Scientific Affairs, Office

of Environmental Policy, Wildlife Trafficking section: http://www.state.
gov/g/oes/env/wlt/index.htm.

13   *every day at U.S. airports:* Again, all examples are real, from sources cited
      above. For more about trends, or the scale and changing nature of wildlife
      crime, see: Liana Sun Wyler and Pervaze A. Sheikh, "International
      illegal trade in wildlife: Threats and U.S. policy," Congressional Research
      Service, March 5, 2008; Jolene Lin, "Tackling Southeast Asia's illegal
      wildlife trade," *Singapore Year Book of International Law,* no. 9 (2005):
      191–208; Dilys Roe, Teresa Mulliken, Simon Milledge, and Josephine
      Mremi, "Making a killing or making a living? Wildlife trade, trade
      controls and rural livelihoods," International Institute for Environment
      and Development, 2002; and Mara E. Zimmerman, "The black market for
      wildlife: Combating transnational organized crime in the illegal wildlife
      trade," *Vanderbilt Journal of Transnational Law* 36, no. 1657 (2003): 657.

16   *Like narcotics officers nabbing street dealers:* Ed Volz, interview with
      author; Kevin Harrington, interview with author.

17   *"He knows the product you are bringing him":* Dave Ferguson, transcript of
      taped interview with detectives Bill Jarmon and Ed Volz, June 17, 1994.

18   *An anonymous tip had led Volz to Ferguson:* Details of Ferguson's arrest
      come from: J. Janca and Larry Baker, statement of probable cause,
      Washington Department of Fisheries Patrol, March 10, 1994; Dave
      Ferguson, witness statement, March 16, 1994; Janetta E. Sheehan,
      deputy prosecutor for Clallam County, Washington, letter of cooperation
      and immunity, March 25, 1994; Ed Volz to Captain Ron Swatfigure,
      memo, Washington Department of Fish and Wildlife, August 10, 1994.
      Ferguson's criminal record is outlined by Special Agent Dali Borden in
      USA vs. Tak Sum Ho, Jeffrey Scott Jolibois, and John J. Easterbrook, Jr. (CR-
      96-5031), criminal complaint, February 28, 1996.

21   *He had confessed to stealing:* Details about how much abalone Ferguson
      poached and its ecological impact came from interviews with Ed Volz;
      field surveys and a report by biologists Dwight Herren and Don Rothaus
      in April 1994; and an undated presentation prepared by Rothaus.

21   *Ferguson returned to the Chinese restaurant:* From surveillance videos;
      J. Cook, narrative report, Washington Department of Fish and Wildlife,
      August 1, 1994; Volz to prosecutors, memo, July 29, 1994.

22   *"the opportunity exists to have a huge impact":* Volz to Captain Ron

Swatfigure, Washington Department of Fish and Wildlife, memo, August 10, 1994.

23   *Ferguson's girlfriend screamed:* Volz, interview with author; Volz to Swatfigure, memo, September 14, 1994; incident summary, Port Townsend Fire Department, September 11, 1994; Washington Boat Accident Report, Jefferson County Sheriff's Department, September 11, 1994; follow-up report, Jefferson County Sheriff's Department, October 12, 1994.

24   *That fall and winter, Ferguson went back undercover:* Ferguson's work undercover is outlined in weekly (sometimes daily) memos written by Volz to prosecutors from early spring 1994 to March 1995.

CHAPTER TWO: LARGER THAN LIFE

27   *on a sunny day five months later:* Descriptions of Doug Tobin's first meeting with Dennis Lucia, including such details as the manager moving about behind while they spoke, came from a series of interviews with Lucia.

28   *Gray whales, frequent visitors to Puget Sound:* See: Linda Woo, "'There He Is. He's Cool.' Gray Whale Charms Admirers Along Purdy Spit," *News Tribune,* April 6, 1995.

28   *Doug Tobin had always been big:* Descriptions of Tobin come primarily from my interactions with him. The direct quotes he used during interviews with me. His hair was gray when I met him, but the dark curled-up ringlets can be seen in earlier photographs and surveillance footage. Similarities to Louis XIV are most noticeable in a photograph taken by federal agents in 1996. Other details came from: Detective Paul Buerger, interview with author; Nancy Lyles, interview with author and an April 8, 1987, letter she wrote to the judge before Tobin's sentencing on manslaughter charges; Tobin's fisherman friend Bob Newman, letter to court, undated; Doug's mother, Murray Tobin, letter to court, April 21, 1987; Steve Sigo, interview with author; other friends who were interviewed by the author.

31   *"Nothing can exceed the beauty of these waters":* See: *Narrative of the United States Exploring Expedition During the Years 1838, 1839, 1840, 1841 and 1842,* vol. 4, chapter 9, pp. 325–326. Full quote: "Nothing can exceed the beauty of these waters, and their safety: not a shoal exists within the

Straits of Juan de Fuca, Admiralty Inlet, Puget Sound, or Hood's Canal, that can in any way interrupt their navigation by a seventy-four gun ship. I venture nothing in saying there is no country in the world that possesses waters equal to these."

31   *The Sound is awash in the beautiful and the bizarre:* The most amazing details about fish in Puget Sound were culled from the University of Washington's "Key to the Fishes of Puget Sound," a database of fish species compiled by Shannon DeVaney and Theodore W. Pietsch, and hosted by the University of Washington Fish Collection at the Burke Museum of Natural History and Culture.

32   *The geoduck is a Pacific Northwest celebrity:* Quote from the boutique seafood dealer is from the Web site of fish wholesaler Marx Foods: http://www.marxfoods.com.

32   *cable television show host tweaked:* Mike Rowe, host of the Discovery Channel's *Dirty Jobs,* July 18, 2006.

32   *"God does have a sense of humor":* *Chicago Tribune,* July 20, 1989.

32   *Such attention can seem excessive:* C. Lynn Goodwin, Brent Vadopalas, Alex Bradbury, Don Rothaus, interviews with author.

33   *Geoducks produce growth rings:* Are Strom et al., "Preserving low-frequency climate signals in growth records of geoduck clams (*Panopea abrupta*)," *Palaeogeography, Palaeoclimatology, Palaeoecology* 228, nos. 1–2 (2005): 167–178; Are Strom et al., "North Pacific climate recorded in growth rings of geoduck clams: A new tool for paleoenvironmental reconstruction," *Geophysical Research Letters* 31, no. 6 (2004).

33   *In a speech during the faculty's founding retreat:* Quotes from the botany instructor are from the Emmy-nominated documentary *Three Feet Under: Digging Deep for the Geoduck Clam,* written and directed by Seattle native Justin Bookey, 2003.

34   *Through the pinking gray of dawn:* Geoduck harvest descriptions come from tagging along on several trips with geoduck divers between 2003 and 2008.

35   *When it came to training divers:* Lucia, interview with author.

36   *That environment can kill:* Details of diving accidents came from fatality records, Occupational Safety and Health Administration, in the deaths of Leo Blanchette, James Plaskett, and Todd Buckley, 1988–1990.

36   *A gray whale once nosed:* Mark Mikkelsen, interview with author. Other

interactions between divers and sea life came from additional interviews with a least a dozen geoduck divers.

37   *Almost immediately, Doug said, he saw corruption:* Doug Tobin, interview with author. The quote about dive shops was said to me, during an interview in the Pierce County Jail in 2003. Details of the crimes Tobin said that he saw are outlined by Borden, in a case-file memo, July 12, 1996.

41   *the Sound's waters had provided a seafood bounty:* For details about the declining health of Puget Sound, see annual reports of the Puget Sound Partnership, a state government agency tasked with cleaning up the Sound. I reported interactions between steelhead and sea lions for this story: "Northwest Sea Lions Teach Humans the Folly of Fighting Mother Nature," *Pacific Northwest,* Sunday magazine of the *Seattle Times,* September 7, 2008.

42   *They heard poachers cut hull sections from their boats:* Details of the crimes the detectives were seeing came from several dozen case files from 1994 to 1996.

CHAPTER THREE: CLAM KINGS

44   *It began with a voyage in 1838:* General information about the Wilkes Expedition comes from *Narrative of the United States Exploring Expedition During the Years 1838, 1839, 1840, 1841 and 1842,* vol. 4, pp. 300–564. Melville drawing upon Wilkes for *Moby-Dick* is from: Nathaniel Philbrick, *Sea of Glory* (New York: Penguin Books, 2003). Description of specimen collections and subsequent theft is from Richard I. Johnson's five-volume *Papers on Mollusca, 1942–2003,* Museum of Comparative Zoology, Mollusk Department, Harvard University; from Harley Harris Bartlett, "The Reports of the Wilkes Expedition, and the Work of the Specialists in Science," *Proceedings of the American Philosophical Society* 82, no. 5 (June 29, 1940): 650–655; and from William H. Dall, "Some American Conchologists," *Proc. Biol. Soc. Wash.* 4 (1888): 95–134. Correspondence between Joseph Couthouy and President Andrew Jackson from Serial Set, vol. no. 327, Session, vol. no. 7, 25th Congress, 2nd Session, H.Doc 147, February 7, 1838, pp. 47–247. For specific discussions of *Panopea generosa* see: Augustus Addison Gould, "Introduction to the molluscs of the exploring expedition," *Narrative* 12 (1857): introduction and pp. 385–386.

47   *The geoduck plucked from the mudflats at Nisqually:* Until recently, *Panopea generosa,* the clam first found in the Nisqually Delta and described by Gould, and *Panopea abrupta,* a pair of fossilized shells found near the Columbia River the same year and described by another scientist in 1849, were thought to be the same animal. For years, the preferred name among scientists for the geoduck has been the earlier of the two monikers: *Panopea abrupta.* Struggling to trace the geoduck through literature, I sought the help of shellfish biologist Brent Vadopalas. I kept getting confused about which clam was really the *first* found by scientists. He took it upon himself to figure it out. After his own investigation, Vadopalas wound up proving that *Panopea abrupta* was in fact not even a geoduck at all, but a completely different fossil species. The resulting scientific paper, "The proper name for the geoduck: Resurrection of *Panopea generosa* (Gould, 1850), from the synonymy of *Panopea abrupta* (Conrad, 1849)," by Brent Vadopalas, Theodore W. Pietsch, and Carolyn S. Friedman, has been accepted for publication in the winter of 2010 in the journal *Malacologia.* It returns *Panopea generosa* to its place as the true name for the geoduck.

47   *Shellfish had been central:* Curtis W. Marean, M. Bar-Matthews, J. Bernatchez, et al., "Early human use of marine resources and pigment in South Africa during the Middle Pleistocene," *Nature* 449 (2007): 905–908; John Noble Wilford, "Key Human Traits Tied to Shellfish Remains," *New York Times,* October 18, 2007.

47   *From the opening of the Oregon Territories:* "Acres of Clams" was written by Tacoma judge Francis D. Henry, and its lyrics are printed on Ivar's menus. For more on Olympia oysters see: Rowan Jacobsen, *The Living Shore: Rediscovering a Lost World* (New York: Bloomsbury USA, 2009).

48   *One column of a geoduck's siphon:* Vadopalas, interview with author.

49   *"Its flesh is, I think, the most delicious":* Hemphill quote in John A. Ryder, "The geoduck," *Scientific American,* April 29, 1882.

49   *Skillfully cooked:* R. E. C. Stearns, "The edible clams of the Pacific coast," *Bulletin of the United States Fish Commission,* vol. 3, no. 23, October 19, 1883.

49   *Geoducks reached epicurean heights:* untitled, *New York Times,* February 23, 1883; and "The Ichthyophagous Dinner," *New York Times,* October 1, 1884.

49   *a shellfish's popularity sometimes came at a price:* Manier quote from "County Game Commissioner Submits Plan to Darwin," *Morning Olympian,* July 7, 1916.

50 *It would take a folksinger:* Ivar's life history comes from: Dave Stephens, *Ivar: The Life and Times of Ivar Haglund* (Seattle: Dunhill Publishing, 1988), pp. 51–69, 119–123; Seattle historian Paul Dorpat, interview with author; Bob Donegan, president of Ivar's, interview with author; Virginia Kraft, "A High Time on a Low Tide," *Sports Illustrated,* December 14, 1964.

51 *During one excavation in 1960:* The discovery of geoducks by Robert Sheats and the subsequent efforts to establish a clam-fishing industry are detailed in Robert Sheats, letter to Cedric Lindsay, Washington Department of Fisheries, January 5, 1967; Sheats, letter to John A. Biggs, Director, Department of Ecology, December 1, 1971; Sheats correspondence, letter to Eric F. Hurlburt, Washington Department of Fish and Wildlife, June 16, 1981; Sheats, unpublished essay, 1977; Sheats family photographs, 1970; Verda Averill, "Poulsbo Family Begins First Commercial Harvest of Geoducks," *Kitsap County Herald,* June 3, 1970.

52 *Sheats loved this stretch of water:* Robert C. Sheats, *One Man's War: Diving as a Guest of the Emperor, 1942* (Flagstaff, AZ: Best Publishing Co., 1998); and "Project Sealab II Report: An Experimental 45-Day Undersea Saturation Dive at 205 Feet," U.S. Office of Naval Research, 1967.

53 *Sheats's discovery might have remained:* Brian Hodgson's early work with clams came from: Don Webster, interview with author and deposition, Kitsap County Superior Court, November 25, 1975; Jerry Elfendahl, interview with author; Carl Sheats, interview with author; Brian Hodgson, testimony at trial, *State of Washington vs. Rod Carew,* 1977; "Management Plan for the Puget Sound Commercial Geoduck Fishery," State of Washington, September 1981; "Which Shall It Be: Geoduck or King Clam?" *Seattle Times,* September 13, 1970; James Bylin, "Lovers of Shellfish Take Heart! Geoduck, or Gooeyduck, Is Here," *Wall Street Journal,* August 3, 1970; David Miyauchi, Max Patashnik, and George Kudo, "Fish protein used to bind pieces of minced geoduck," *Proceedings of the National Shellfisheries Association* 63 (1973); Hodgson himself told the story about driving over geoducks in his truck in: Susan Herrmann Loomis, "The State's 'King Clam' Has Been a Northwest Regional Delicacy for Centuries," *Seattle Times,* May 26, 1985. Quotes from Hodgson are from the same story.

54 *Hodgson asked one of his best customers:* Shiro Kashiba, interview with author; Webster, interview with author.

55   *Like few other civilizations, the Chinese express:* Most helpful in
      understanding Chinese banquet foods were: Quentin Fong, interview with
      author; Harry Yoshimura, Mutual Fish Company, interview with author;
      Mark Wen, interview with author; and, of course, Claude Tchao, whose
      tale of geoduck discovery was confirmed by some of his competitors.
      Also helpful were Tony Wong, whose father was the other Vancouver
      geoduck entrepreneur, and an e-mail exchange with Lawrence Lai, in
      Hong Kong. See also: Thomas Liu, "An Insight into the Shanghai Market
      for Imported Live Seafood," World Ocean Trading Co. Ltd., Shanghai,
      Marketing and Shipping Live Aquatic Products 221, University of Alaska
      Sea Grant, January 3, 2001; LaVerne E. Brabant, "Peoples Republic of
      China Exporter Guide, 1999," U.S. Consulate General, Shanghai, for
      Foreign Agricultural Service Global Agriculture Information Network,
      U.S. Department of Agriculture, December 15, 1999. For more on the
      role of Chinese food through history see: K. C. Chang, *Food in Chinese
      Culture: Anthropological and Historical Perspectives* (New Haven, CT: Yale
      University Press, 1977); Fuchsia Dunlop, *Shark's Fin and Sichuan Pepper:
      A Memoir of Eating in China* (New York: W. W. Norton, 2008); Jennifer
      8. Lee, *The Fortune Cookie Chronicles: Adventures in the World of Chinese
      Food* (New York: Twelve Books, 2008).

57   *Between 1983 and 1993:* Lawrence Lai, "Marine fish production and
      marketing for a Chinese food market: A transaction cost perspective,"
      Aquaculture Economics & Management, September 2005, pp. 289–316.

57   *The number of toy factories:* "A Day in the Life . . . of China: Free to Fly
      Inside the Cage," *Time*, October 2, 1989.

57   *geoduck fishing remained tightly controlled:* For more anecdotes about
      Hodgson breaking the rules see: *State of Washington vs. Brian Hodgson*,
      Case No. 88-1-05697-4, amended certification of probable cause.

58   *his divers told a fussy regulator:* Tom Gillick, former shellfish program
      manager, Department of Natural Resources, memorandum of interview
      with Assistant Attorney General Jay Geck, April 28, 1989.

58   *In June 1987, a band of protesters:* Dick Clever, "Clamscam: Media Got
      Conned," *Seattle Times*, June 11, 1987; and "ClamScam: One Shell of
      a Hoax," *New York Post*, June 11, 1987. For more on the practical joker
      himself see: http://www.joeyskaggs.com.

59   *Rumors about Hodgson overfishing:* Dee Norton, "Giant Clam Industry

Scrutinized by Feds," *Seattle Times,* February 16, 1982; Rod Carew,
testimony, federal grand jury, U.S. District Court, Seattle, January
19, 1982; Rod Carew, testimony, state court of inquiry, 1987; Marilyn
Brenneman, interview with author and copies of her written opening
statement in trial; Detective Kevin Harrington, interview with author;
Clifford Bergerson, enforcement report, January 4, 1987.

62   *Before Hodgson was sentenced, prankster Joey Skaggs:* Joey Skaggs, letter
to State Senator Mike Kreidler, March 13, 1989.

CHAPTER FOUR: THE FED

63   *Kevin Harrington shot south:* Details of the meeting between Harrington,
Severtson, Borden, and Tobin are outlined in Borden's July 12, 1996,
case-file memo. Details about Severtson's reaction to Tobin are from
Harrington and Severtson, in interviews with the author. Details about
the potential suspects come from copies of the flow chart.

65   *This seafood crime wave was gathering steam:* Keith B. Richburg, "China
to Buttress Hong Kong Police—But to What End?" *Washington Post,* May
4, 1997; several stories in *Wen Wei Po,* a Chinese-language newspaper in
Hong Kong, detailed current and past seafood smuggling operations from
Hong Kong to Kat O. The stories ran December 8, 2003; April 10, 2005;
and November 10, 2005. Similar reports appeared in *South China Daily* in
July 2002; and in reports by Guangdong Province's Oceanic and Fisheries
Administrator. All were translated for me by freelance writer Ting-I Tsai.
Geoduck smuggling clearly continues, as evidenced by this: "Sub-bureau
of Anti-Smuggling of Panyu Customs House Intercepted and Captured
Smuggled Seafood," press release, Guangzhou Customs District, People's
Republic of China, March 16, 2007.

65   *a half-billion dollars in American wildlife:* Timothy J. Larsen, "The
Chinese market for Colorado and the U.S. agricultural exports:
Analysis of the potential impacts of the establishment of permanent
normal trade relations (PNTR) with China and Colorado's agricultural
industry," Colorado Department of Agriculture, September 28, 2000;
Joe Haberstroh, "China Cuts Tariff on Apples—Move Will Mean
More Exports for State Growers," *Seattle Times*, January 14, 1994; Joe
Haberstroh, "Ripe for the Exporting—State Industry Eyes a Possibly
Fruitful China Market with Growing Anticipation, Caution," *Seattle*

Times, February 6, 1994; Jennifer Hieger, "Apple Growers Could Reap Fruits of China's WTO Entry," *Yakima Herald-Republic,* April 16, 1999.

69　　*Volz and Harrington had worked with Severtson before:* Volz and Harrington, interviews with author.

69　　*But Severtson could also alienate his colleagues:* Andy Cohen, Al Samuels, Wayne Lewis, Volz, and Harrington, interviews with author.

70　　*Severtson's disdain for administrators:* Wayne C. Lewis, *Sea Cop* (Oregon: River Graphics, 2004), pp. 109–110.

70　　*But his investigative prowess was legendary:* Ibid., pp. 164–179; Severtson, interview; Lewis, interview; Cohen, interview; Hal Bernton, "Cargo Ship's 'Milk Run' Becomes an Adventure," *Anchorage Daily News,* July 27, 1989.

72　　*typed his reports in his own small office:* Samuels, interview; Cohen, interview.

74　　*Severtson had great confidence:* Severtson, interview.

CHAPTER FIVE: METAMORPHOSIS: LIFE UNDERCOVER

76　　*Sell a dozen San Francisco garter snakes:* Sam Jojola, interview with author.

76　　*Supplying medical researchers with primates: USA vs. LABS of Va. Inc.* (02-CR-0312). More than one thousand crab-eating macaques had been shipped out of Indonesia over several years, some just a few weeks old.

76　　*Newcomer began figuring all this out:* Entire investigative file, World Insect, Inc. (INV: 2003102546), and court file, *USA vs. Kojima* (06-CR-00595), including: narrative reports, audio recordings of undercover conversations, undercover video, and copies of Skype video conferences. All quotes and conversations with Hisayoshi Kojima are verbatim from audio and video recordings, except where noted. Information also came from: Ed Newcomer, interview with author; Jojola, interview; John Brooks, interview; John Gavitt, interview; Jerry Smith, interview; Ernest Mayer, interview; Erin Dean, interview; Lisa Nichols, interview; Skip Wissinger, interview. Descriptions of bug fair come from videotapes of earlier fairs, advertisements, and from a brief undercover video shot by Newcomer.

83　　*A long line of men and women:* See: "Three U. S. game and law agents retire from active service," Department of Interior, September 30, 1940; Louis S. Warren, *The Hunter's Game: Poachers and Conservationists in Twentieth-Century America* (New Haven, CT: Yale University Press, 1997), pp. 34–43; and "Wildlife Bureau Agents Are Competent Detectives," *Hartford Courant,* October 26, 1941.

83   *Peanut Man dispensed nuts:* Jerry Smith, interview.

84   *The Yoshi files sat untouched:* Newcomer, interview; and in probable cause affadavit for the roller pigeon case, filed by Newcomer in *USA vs. McGhee et al.* (07-CR-0737), May 17, 2007.

85   *"What are you doing here?":* This exchange is as recollected by Newcomer.

86   *"You ever get these chimaeras?":* Conversation is verbatim, from audio recording.

91   *In a recorded telephone call:* Quotes here are taken from Newcomer's affadavit, where they are recorded as direct quotes.

91   *Newcomer brought Yoshi down:* All quotes here are taken directly from copies of video chats between Newcomer and Kojima.

92   *Tobin shared Newcomer's talents:* From Severtson, Volz, Samuels, interviews with author.

92   *Severtson wanted Tobin to get close to Gene Canfield:* Quotes in the opening paragraph are from Canfield, interview with author. All remaining quotes are from transcripts of recorded conversations between Canfield and Tobin.

93   *Canfield had told Tobin about a California clam market:* Gene Canfield's role in the DeCourville case was detailed by Charles Tyer in a search warrant affadavit, *USA vs. Premises Known as 4865 Rollingwood* (CR-97-1107), June 18, 1997; defendant's sentencing memorandum, *USA vs. DeCourville* (CR97-5301), December 17, 1997; daily memos to file by agents Severtson and Dali Borden documenting illegal sales; memo to Micki Brunner and J. Lord from Dali Borden, October 7, 1997; and transcripts and official synopses of undercover telephone calls.

93   *Late on a hot August night in 1996:* Transcript of undercover call between Tobin and Canfield, 10:30 P.M., August 21, 1996.

94   *"In China?" Canfield asked*: Sheryl WuDunn, "China's Rush to Riches," *New York Times Magazine,* September 4, 1994; Wen Wei Po stories; Severtson, interview with author.

94   *Canfield tutored him in illegal geoduck harvesting:* Tape synopsis of

undercover meeting between Tobin and Canfield, August 25, 1996; search warrant in *USA vs. Premises Known as 4865 Rollingwood.*

95   *The next day, Severtson opened an account:* Severtson, memos to file. Copies of invoices used in these transactions for Tobin's company, Blue Raven, which depict a bird in flight. The logo was designed by Special Agent Al Samuels.

95   *That night, Severtson, Borden, and Bill Jarmon:* Severtson memo to file, "Sale of Product on August 27, 1996," September 30, 1996; defendant's sentencing memorandum, *USA vs. DeCourville.*

96   *The next day the Las Vegas broker called Tobin:* Sentencing memorandum, *USA vs. DeCourville.*

CHAPTER SIX: KINGPIN

97   *The detectives knew the name:* Volz, memo to file, synopsis of contacts with Jong Park by statewide investigative unit, May 17, 1996; David Pearson, witness interview with Volz and Harrington, September 18, 1996; Jong Park, voluntary statement, Department of Fish and Wildlife, January 24, 1997; Steve Hoidal, witness statement to Severtson and Volz, September 23, 1997.

98   *DeCourville had contacted the cops:* Tyer, search warrant affadavit, *USA vs. Premises Known as 4865 Rollingwood.*

98   *The investigators learned about DeCourville:* Most profile information taken from *USA vs. DeCourville,* defendant's sentencing memorandum, December 17, 1997.

99   *DeCourville eventually worked his way into:* Beverly Beyette, "Hefner and Ex-Bunnies Set to Toast the End of an Era," *Los Angeles Times,* June 30, 1986; James Bartels, interview with author.

99   *One DeCourville patron:* Isgro's status as a regular came from Bartels. Isgro's legal issues are detailed in: *USA vs. Isgro* (CR-89-00951) and *USA vs. Isgro* (CR-00-00326).

99   *By then DeCourville had remarried:* Linda Gentille, interview with author; Mary Barber, "Sit-down Comic at the Keyboard Plays in Some Strange Situations," *Los Angeles Times,* January 17, 1985; Carrie Delmar, "She Keyes Elderly to High Note," *Los Angeles Daily News,* August 13, 1986.

100   *Gene Canfield and Doug Tobin kept talking regularly:* Charles Tyer, search warrant affadavit, *USA vs. Premises Known as 4865 Rollingwood;*

defendant's sentencing memorandum, *USA vs. DeCourville;* daily memos to file by agents Severtson and Dali Borden, August–October 1996; and in transcripts and official synopses of undercover telephone calls.

100   *The Vegas broker quickly warmed to his new:* This conversation was recorded by Borden on September 11, 1996. Volz played a cassette copy of it for me in spring 2008.

101   *Canfield would "never rat anybody out":* Quotes taken from transcript of telephone call recorded on September 5, 1996.

102   *DeCourville confided that he controlled 70 percent:* Transcript of call, September 5, 1996.

102   *By early fall, though, the state:* Borden, Severtson, in dozens of memos to file, August–December 1996.

102   *Sometimes it was the rookie federal agent:* Samuels, interview; Rothaus, interview.

103   *DeCourville remained confident he could outsmart mere fish cops:* Quotes from transcript of undercover call, September 25, 1996.

103   *Federal agent Vicki Nomura began posing:* Quotes from transcript, October 1, 1996.

104   *DeCourville had been getting divers to "pencil whip" the harvest:* "Paralytic shellfish poisoning strikes Kodiak," *Epidemiology Bulletin,* State of Alaska, no. 13, June 20, 1994; "Epidemiologic notes and reports: Shellfish poisoning—Massachussets and Alaska, 1990," *Morbidity and Mortality Weekly Report,* Centers for Disease Control, March 15, 1991. For more on dinoflagellates see: Edward R. Ricciuti, *Killers of the Seas: The Dangerous Creatures that Threaten Man in an Alien Environment* (Guilford, CT: Lyons Press, 2003), pp. 143–148.

106   *But the risk isn't just to the other side of the world:* The story of the midwestern monkeypox outbreak was shared with me most eloquently by Darin Carroll, interview with author; and Jennifer McQuiston, interview. For more information see the CDC's monkeypox page: http:// www.cdc.gov/ncidod/monkeypox/; and Liana Sun Wyler and Pervase Sheikh, "International illegal trade in wildlife: Threats and U.S. policy," Congressional Research Service, updated August 2008. See also: Margaret Ebrahim, "Giant Rats Spread Disease in Florida," Associated Press, November 29, 2006.

108   *some investigators remained skeptical:* Volz, Harrington, Samuels, interviews.

109   *The long history of antagonism:* Accounts of the Northwest fishing wars
      of the mid-twentieth century are culled from interviews with Volz,
      Jarmon, Buerger, and Tobin; from a series of stories written by Robert
      Mottram in the *News Tribune:* "Angry Gillnetters Ram 7 Patrol Boats,"
      October 7, 1976; "FBI Fishing Violence Probe Sought," October 21, 1976;
      "Shot Fisherman Paralyzed," October 28, 1976; from Janet McCloud and
      Robert Casey, "The Last Indian War," in the monthly Indian newsletter
      Survival News, Seattle bulletins nos. 29 and 30; and from several books:
      *Uncommon Controversy: Fishing Rights of the Muckleshoot, Puyallup and
      Nisqually Indians* (Seattle: University of Washington Press, 1970); Fay
      G. Cohen, *Treaties on Trial: The Continuing Controversy over Northwest
      Indian Fishing Rights* (Seattle: University of Washington Press, 1986); and
      Charles Wilkinson, *Messages from Frank's Landing* (Seattle: University of
      Washington Press, 2000).

111   *Severtson tried in subtle ways to keep Tobin in check:* Severtson, Samuels,
      interviews.

112   *The cops' previous informant, Dave Ferguson:* Gene Porter and Richard
      Hansen, *USA vs. Jeffrey Jolibois* (96-CR-5320), evidentiary hearing,
      sentencing, November 15, 1996.

113   *"Where's the monitor?":* Quotes from tape synopsis, May 12, 1997.

114   *Prices fluctuated by the day:* Evelyn Iritani, "Geoducks: Garbage into
      Gold," *Los Angeles Times,* April 7, 1997.

CHAPTER SEVEN: "IT'S JUST A BUSINESS THING"

116   *Wildlife trafficking attracted violence:* See source list from "Elephant
      Tusks, Wild Furs" in chapter one. See also: Dee Cook, Martin Roberts,
      and Jason Lowther, "The international wildlife trade and organised crime:
      A review of the evidence and the role of the UK," Regional Research
      Institute, University of Wolverhampton, June 2002; David E. Kaplan and
      Alec Dubro, *Yakuza: Japan's Criminal Underworld* (Berkeley: University
      of California Press, 2003).

117   *In 1991 Brooklyn police found:* Thomas Dades, interview.

117   *In the 1980s, Special Agent Andy Cohen saw:* Cohen, interview; *Sea Cop,*
      pp. 174–175.

118   *Tobin arrived twenty minutes late:* Severtson, interview.

118   *"That's not right":* Quotes from transcript of call placed at 9:20 P.M., June 4, 1997.

120   *Bakker, a fast-talking broker:* Severtson, Samuels, Casey Bakker, interviews.

121   *"They're going to call me back tonight":* Severtson played the actual audio recording of this call for me. I also have the transcript. It was placed at 4:45 P.M., June 5, 1997.

124   *Assistant U.S. Attorney Helen "Micki" Brunner:* Micki Brunner, interview; quote is as recalled by Brunner.

125   *Severtson checked in on agents Cohen and Samuels:* Cohen, Samuels, interviews.

125   *the agents were certain they'd found their man:* Rick Jones is mentioned throughout *USA vs. Nichols DeCourville* (97-CR-5056), complaint for violation, August 27, 1997; by James Bartels in memorandum of interview with Severtson, July 1, 1997.

125   *On a Sunday in June 1997, Jones rolled out:* This section is taken from an FBI 302 memo by Special Agents R. T. Ballard and Stanley E. Orenallas, from their interview with Jones on June 19, 1997; from Severtson's July 1 interview with Bartels; from copies of the Best Western motel bill with Jones's name on it; from a report to file of Severtson and Cohen's interview with Jones, July 2, 1997; from a report of Severtson's interview with DeCourville on June 19, 1997; from telephone records; and from my interview with Bartels.

130   *A phalanx of state and federal agents:* This scene is culled primarily from: Vicki Nomura, report to file, arrest and search warrants in Las Vegas, July 8, 1997. Nomura wrote thirteen single-spaced pages about the day's events, and her report includes details about DeCourville's facial tics and the level of smoke in the air. Also used were: Severtson, execution of search and arrest, memo to file, June 24, 1997; Volz, search warrant service at DeCourville residence, June 25, 1997; and interviews with Volz, Severtson, and Tyer.

130   *Ron Peregrin, had bought wraparound casts:* Ron Peregrin, interview. Bakker doesn't recall the details but doesn't dispute Peregrin's recollection.

132   *At 6:35 A.M., Detective Kevin Harrington:* From reports by Fish and

Wildlife officers Ron Druer and Larry Baker and a minute-by-minute search execution log by Ralph Woods, all dated June 19, 1997; and from a narrative report by Harrington, June 20, 1997.

133 *That same morning, a caravan:* Peregrin, interview; search warrant affadavit, unsealed at my request.

134 *Also that day, two FBI agents:* From Ballard and Orenallas, 302 memo.

135 *Two weeks after the raids:* Severtson and Cohen, report to file, interview with Jones, July 2, 1997; Cohen, interview; I have a photocopy of the National Marine Fisheries Service photograph of Jones holding his stuffed animal.

CHAPTER EIGHT: AN INCREDIBLE VIRUS

137 *Flipping through the DeCourville files:* From Brunner, interview; Cohen, interview; Buerger, interview; Brunner's opening and closing statements in *USA vs. Hansen-Sturm* (93-CR-02166), 1993; Stephen Darnell, memorandum of interview with Cohen, 1992; U.S. Customs, report of investigation, Hansen Caviar Company, February 15, 1991; Cohen, affadavit for search warrant, Hansen Caviar Company, June 15, 1992; Cohen, "Sturgeon Poaching and Black Market Caviar: A Case Study," *Underwater Naturalist* 28, no. 1 (2004): 23–26; and Lee Wohlfert, "That Spoonful of Fish Eggs Costs a Mere $12," *People,* August 21, 1978.

140 *The DeCourville case could also make headlines:* Brunner, interview; Rick Anderson, "Assailant Paroled; Brother Still in Prison," *Seattle Times,* December 6, 1977; Paul Clegg, "20-Year Sentence for Attack on Former Footballer," *Bremerton Sun,* April 7, 1976; *State of Washington vs. Douglas John Martin Tobin* (86-1-342-1), entire case file; *State of Washington vs. Daryl Burns and David Jirovec* (86-1-366-9), entire case file.

143 *Tobin's story sounded implausible:* Severtson, interview; Harrington, interview; Volz, interview; Jarmon, interview; Cohen, interview; Samuels, interview. Several investigative case files.

147 *Doug Tobin, on the other hand:* Tobin's artwork has been well displayed in galleries in Olympia. Jay Geck, a lawyer with the state attorney general's office, saw Tobin and his work on display during a fund-raising event for a local land trust. See also: Marsha King, "New Project to Showcase NW Native Art—Plan Focuses on Tribal Self-Sufficiency," *Seattle Times,* August 13, 1996. Information about the welcome pole is from:

Tobin, interview; Duane Pasco, interview; Edward Binder, interview; Robin Wright, interview; Russell Lidman and Michael Bisesi, "The Welcome Pole: Public Art, Process, and Controversy," *The Journal of Arts Management, Law and Society* 34, no. 4 (Winter 2005): 245–261.

148 *Tobin seemed equally triumphant in business:* I had two short interviews with Jeffrey Albulet, who told me the story about Tobin helping the homeless man. Other details about his relationship with Albulet and Ng come from *State of Washington vs. Albulet and Ng* (05-2-254-6), and include affadavits from several geoduck divers; a sixteen-page single-spaced memorandum of an interview with Doug Tobin by Karolyn R. Klohe, assistant attorney general, June 15, 2005; transcripts of attorney general interviews with several additional members of Tobin's crew.

149 *Tobin partnered with a friend, Adrian Lugo:* Details are from Adrian Lugo, interview; transcripts of an attorney general interview with Adrian Lugo, April 13, 2005; Himanee Gupta, "Entrepreneur Fulfills Artistic Interests through Construction," *Seattle Times,* April 22, 1991; "Brief profiles of 15 Inc. 500 companies and their founders," *Inc.,* December 1, 1990. Lugo's company was number 274 on the magazine's list of top 500 small businesses. Details about the tractor salesman come from Volz, interview; a timeline of contacts between wildlife agents, Gary Ufer, and Doug Tobin in 1997; a witness statement from Tobin about Ufer to Severtson and Volz, October 15, 1998; and a statement from Ufer about Tobin to detectives, October 21, 1998.

151 *Ed Volz could feel the Feds' priorities:* Volz, Severtson, interviews. Paul Watson, interview. For more about the whale hunt see: Robert Sullivan, *A Whale Hunt: Two Years on the Olympic Peninsula with the Makah and Their Canoe* (New York: Simon & Schuster, 2000).

155 *On a cold morning early in 2000:* Descriptions are all taken from the video footage itself.

CHAPTER NINE: A SEA OF ABUNDANCE

159 *Puget Sound holds millions of geoducks:* Vadopalas, Bradbury, Goodwin, Sizemore, Rothaus, Pauly, interviews.

160 *History is filled with once-prolific sea creatures:* From Claudio Richter, Hilly Roa-Quiaoit, Carin Jantzen, Mohammad Al-Zibdah, and Marc Kochzius, "Collapse of a new living species of giant clam in the Red Sea," *Current*

Biology 18, no. 17 (September 9, 2008): 1349–1354. Interviews and e-mail exchanges with Daniel Pauly and fellow marine biologist Jennifer Jacquet helped me grasp this study's significance.

161   *Stocks of the world's most prized fish:* I gathered some of this information with a colleague for a story in 2004 about John Kerry. See: Hal Bernton and Craig Welch, "Fight over Fishing Tested Kerry's 'Green' Credentials," *Seattle Times,* September 30, 2004. See also: Beth Daley and Gareth Cook, "A Once-Great New England Fishing Industry on Brink, as Rules Squander Catches," *Boston Globe,* October 26, 2003. For the best full accounting of cod's significance and collapse see: Mark Kurlansky, *Cod: A Biography of the Fish That Changed the World* (New York: Penguin, 1997).

161   *Still, invertebrates such as shellfish, sea urchins:* I also reported this for the *Seattle Times*. See: Craig Welch, "Abalone Are Treasured—Nearly to Extinction," *Seattle Times,* May 13, 2009. Also interviews with: Bradbury, Rothaus, Vadopalas.

163   *About the time Ed Volz was making:* Vadopalas, Bradbury, interviews. Bradbury articulated a distinction between extinction and collapse. Even with an ecological catastrophe—which, of course, could come with climate change or ocean acidification—geoducks are extremely unlikely to face *extinction* from fishing or poaching. There are too many of them. But the number of geoducks could drop substantially and drive the clam-fishing industry toward collapse. That's what happened with Atlantic cod. It is by no means gone, just too scarce to be fished at the level it once was. That drives fishermen to other species, changes ocean ecology, and leaves cod populations more vulnerable to disease and environmental changes. For more see: Ray Hilborn and Carl J. Walters, *Quantitative Fisheries Stock Assessment: Choice, Dynamics and Uncertainty* (Norwell, MA: Kluwer Academic Publishers, 1992). See also: J. M. (Lobo) Orensanz, Claudia M. Hand, Ana M. Parma, et al., "Precaution in the harvest of Methuselah's clams—the difficulty of getting timely feedback from slow-paced dynamics," *Canadian Journal of Fisheries and Aquatic Sciencies* 61, no. 8 (2004): 1355–1372; and Juan L. Valero et al., "Geoduck recruitment in the Pacific Northwest," California Cooperative Oceanic Fisheries Investigations Reports, vol. 45, 2004.

165   *There was also the looming threat:* A. Whitman Miller, Amanda C. Reynolds,

et al., "Shellfish face uncertain future in high CO_2 world: Influence of acidification on oyster larvae calcification and growth in estuaries," *PloS One* 4, no. 5 (May 27, 2009). See also: Craig Welch, "Oysters in Deep Trouble: Is Pacific Ocean's Chemistry Killing Sea Life?" *Seattle Times,* June 14, 2009.

166 *By the turn of the century, overfishing and seafood smuggling were commonplace:* Velisarios Kattoulas, "The death of sushi: Japan's passion for sushi is fueling a huge trade in illegally caught seafood that's endangering fish stocks and enriching organized crime," *Far Eastern Economic Review,* August 15, 2002; Elisabeth Rosenthal, "Europe's Appetite for Seafood Propels Illegal Trade," *New York Times,* January 15, 2008. For more about the decline of the marine world's largest ocean fish see: Ransom A. Myers and Boris Worm, "Rapid worldwide depletion of predatory fish communities" *Nature,* May 15, 2003; and Worm, Ray Hilborn, et al., "Rebuilding global fisheries," *Science* 325, no. 5940 (July 31, 2009): 578–585.

167 *The trouble came to light on May 15, 1991:* OSHA records of Howard's death have been destroyed, but accounts of it could be found in: Bill Beebe, "Battle Rages over Collector Fish," *Daily Breeze,* June 4, 1992; and Ronald B. Taylor, "Aquarium Fish Trade Comes under Attack," *Los Angeles Times,* June 21, 1992. Suzanne Kohin, interview with author. Also: Colette Wabnitz, Michelle Taylor, Edmund Green, and Tries Razak, "From ocean to aquarium: A global trade in marine ornamental species," UNEP World Conservation Monitoring Centre, 2003.

169 *special agent Roy Torres heard from a colleague in Miami:* Roy Torres, interview with author. Maureen Bessette, interview with author. Entire investigative file, *Leopard Shark Case* (SW040119), and court file, *USA vs. Thompson* (06-CR-00051), including: narrative reports of Torres's investigation, audio recordings and transcripts of witness interviews, copies of a sermon by Kevin Thompson. All quotes and conversations are verbatim from audio recordings. Information also came from: Lisa Nichols, interview; Rebecca Hartman, interview.

170 *The Reverend Sun Myung Moon:* Moon has several sermons dedicated to the sea, and he published them as a book, *God's Will and the Ocean*, published by the Holy Spirit Association for the Unification of World Christianity, 1987. I quoted from "The Way of the Tuna," which he gave July 13, 1980. To understand his move into fisheries, I read: Edward Quill,

"Unification Church Blesses Its Own Fleet," *Boston Globe,* July 2, 1981; Todd Heath, "Unification Church in Quiet but Edgy Coexistence with Old Fishing Town," Associated Press, July 1, 1983; Carolyn Lumsden, "Moonies Win Grudging Acceptance from Wary Gloucester Fishermen," *Hartford Courant,* May 18, 1986. For more information see: Monica Eng, Delroy Alexander, and David Jackson, "Sushi and Rev. Moon," *Chicago Tribune,* April 11, 2006.

172   *Even to an eighteen-year-old, it seemed obvious:* Quotes in this section are from interviews with suspects conducted by Torres.

175   *Because scientists still don't fully understand the ocean:* Greg Cailliet, Vincent Gallucci, Christopher Lowe, interviews with author.

CHAPTER TEN: CRAB MEN

176   *Geoduck enforcement improved marginally:* There were several stories and editorials in the Tacoma newspaper, the *News Tribune.* One of the earliest reports appeared in the *Central Kitsap Reporter* on April 4, 2000.

176   *They could have used more cameras elsewhere:* Jarmon, interview; Jarmon first heard about the "mystery boat" on June 18, 2000, according to his notes. Bob Mottram, "Mystery Crabbing Boat Draws Concern for South Puget Sound," *News Tribune,* June 25, 2000.

178   *Detective Bill Jarmon's mind was elsewhere:* I met Jarmon in his living room in 2003, and he walked me through the morning. Then I climbed into his Ford Expedition and he drove with me the route he had followed that morning. I carried with me Jarmon's narrative report, from June 28, 2000. I have a photocopy of the citation Jarmon wrote Tobin in 1975.

182   *The detectives had not paid much attention to Tobin:* Tobin's role in pointing out Hodgson appears in: Paul Buerger, memo to intelligence file, October 19, 1999; Pudwill, surveillance detail, October 20, 1999; and Harrington, affadavit for search warrant, undated. Buerger and Harrington would present a case to prosecutors in 2002, but Hodgson would ultimately be aquitted on all charges.

183   *Squaxin Island stretches like a knife handle:* Rory Gilliland, interview with author. Gilliland, field investigation report, Squaxin Island Tribal Enforcement, June 15, 2000; a report of violation, Squaxin Island Police, June 20, 2000; and a supplementary report, June 25, 2000.

185 *Ed Volz growled at the news:* Volz, Charles Amner, and Jarmon, interviews; Harrington, undated search warrant affadavit in Hodgson case; Jarmon, suspect-contact report, October 17, 2000. All quotes from Li and Tobin are taken directly from law-enforcement documents, where they were recorded as direct quotes.

188 *By early spring 2001, word had filtered back:* Volz, Jarmon, Harrington, interviews. Jarmon, report of surveillance, March 8–9, 2001; Pudwill, report of surveillance, March 13–14, 2001.

189 *Their first significant break came with:* Volz, interview. Harrington, report of interviews with Mark Purdue and Heidi Mills, April 24, 2001. Mills, interview with author, 2003.

192 *Frustration mounted that summer:* Volz, report of information from informant, June 21, 2001; Volz, report of information from informant, June 25, 2001; Jarmon, report of contact with Keith Smith, August 24, 2001; Jarmon, summary of interviews with Keith Smith, Ocober 18, 2001.

196 *The magnitude of what Tobin had done sunk in slowly:* Jerry McCourt, interview with author. Copies of McCourt's notes to Jarmon. Reports of surveillance by Ed Volz, Bill Jarmon, and Charlie Pudwill, on October 25, October 30, November 6, and November 8.

CHAPTER ELEVEN: THE HUNT, REDUX

199 *From a pea-gravel beach a few minutes later:* See notes from prologue. Also, interview with Don Rothaus; Rothaus and Michael Ulrich, report of investigation of geoduck tract nos. 12300 and 12350, November 13, 2001. Pudwill, interview. Pudwill, report of surveillance, November 14, 2001.

203 *The detectives regrouped:* In 2008, I watched the video Pudwill shot of Tobin at his plant. Jarmon, Volz, Harrington, Pudwill, interviews. Jarmon, report of surveillance, June 18–19, 2001; Jarmon, report of surveillance, June 20–21, 2001; Pudwill, report of surveillance, June 20, 2001.

204 *A few days later Tobin paged Kevin:* Harrington, interview. Harrington, memo to prosecutor about conversation with Tobin, November 28, 2001. Volz, supplemental report, call from Tobin, December 3, 2001. Quotes here are as Volz recalled them.

206 *In January . . . Tobin was still calling:* The call from Mark Purdue was recorded and transcribed and quotes presented here are verbatim. Harrington, supplemental report, call from Tobin, February 11–12, 2002.

207   *Tobin could not have known:* Volz, report of conversation with Matt
      Donovick, January 8, 2002; February 18, 2002. Harrington, report of
      interview with Donovick, February 19, 2002; February 21, 2002. Volz,
      call from Donovick, March 1, 2002. Jarmon, report of contacts with Heidi
      Mills, March 10, 13–14, 2001.

209   *Early in March, Adrian Lugo:* Lugo, interview with author.

210   *On March 13, 2002, Tobin called Harrington:* Harrington, Jarmon,
      interviews. Harrington, supplemental report, call from Tobin, March 13,
      2002.

CHAPTER TWELVE: THE WHOLE WEST COAST

212   *Between Doug Tobin's calls and information gleaned:* Jarmon, report of
      contact with Heidi Mills, March 17, 2002. Jarmon, Volz, interviews.

213   *At 6:30 A.M., police and detectives moved:* Fife Police Department,
      supplemental narrative, March 18, 2002; Dan Chadwick, officers report
      to prosecutor, March 18, 2002; T. Jackson, supplemental report, March
      18, 2002; Jarmon, incident report, "Operation Typhoon," arrest warrant
      execution, March 18, 2002; Dreher, report of investigation, March 18,
      2002; Pudwill, search warrant service, memo to prosecutor, March
      18, 2002. Video of search at Tobin's plant. Details about Tobin's arrest,
      including Volz holding the phone to call Tobin's lawyer, are included in
      the reports.

214   *The wildlife cops eventually arrested:* William Omaits, forensic accountant,
      affadavit, February 13, 2004; Bob Sizemore, affadavit, February 11, 2004;
      Wayne Palsson, affadavit, January 30, 2004; *State vs. Tobin* (02-1-01236-3),
      entire case file.

216   *Tobin waited month after month in jail:* Quotes direct from transcript of
      recorded telephone call between Tobin and aging poacher.

217   *In court, Doug Tobin crumpled into his seat:* I was in court to witness
      this scene. I saw Volz twirling his toothpick while talking with Tobin's
      attorney. The last quote from Volz was said to me, standing outside the
      courtroom.

219   *A month after Doug Tobin began his incarceration:* This meeting was
      videotaped, and this scene is taken from a copy of that tape.

223   *Eight months later, authorities hauled Tobin:* From entire court file, *State
      vs. Doug Tobin* (04-1-04236-6). Jarmon, Volz, interviews. Fire marshal,

report of investigation. Also from the lawsuit filed against the *Laurie Ann*'s owner, *Pierce County vs. Washington Shellfish, Inc. and Douglas McRae* (02-2-08561-5).

225   *Detective Volz wanted* The Typhoon: Klohe, interview with author; Joe Panesko, interview with author; quotes from Albulet are from first of two short author interviews. Also see: *State of Washington vs. Albulet and Ng* (05-2-254-6), which includes affadavits from several geoduck divers; a sixteen-page single-spaced memorandum of interview with Doug Tobin by Karolyn R. Klohe, assistant attorney general, June 15, 2005; transcripts of attorney general interviews with several additional members of Tobin's crew. The framed check hangs in Panesko's office.

SOME OF THOSE INTERVIEWED FOR THIS PROJECT INCLUDE:
State wildlife cops

Ed Volz, detective, Washington Department of Fish and Wildlife
Bill Jarmon, detective, Washington Department of Fish and Wildlife
Kevin Harrington, detective, Washington Department of Fish and Wildlife
Paul Buerger, detective, Washington Department of Fish and Wildlife
Charlie Pudwill, detective, Washington Department of Fish and Wildlife
Ron Peregrin, undersheriff, Clallam County, former WDFW detective
Nancy Foley, chief of law enforcement, California Fish and Game
Kathy Ponting, special operations, California Fish and Game
Troy Bruce, special operations, California Fish and Game
Rebecca Hartman, enforcement officer, California Fish and Game

Federal agents

Andy Cohen, special agent, NOAA
Al Samuels, special agent, NOAA
Richard Severtson, special agent, NOAA
Roy Torres, special agent, NOAA
Charles Tyer, special agent, NOAA
Wayne Lewis, special agent in charge (retired), NOAA, Seattle
Michelle Zetwo, special agent, NOAA
Ed Newcomer, special agent, U.S. Fish and Wildlife Service
Sam Jojola, special agent (retired), U.S. Fish and Wildlife Service
Lisa Nichols, special agent, U.S. Fish and Wildlife Service

John Gavitt, special agent (retired), U.S. Fish and Wildlife Service
Ernest Mayer, special agent (retired), U.S. Fish and Wildlife Service
Jerry Smith, special agent (retired), U.S. Fish and Wildlife Service
John Brooks, special agent (retired), U.S. Fish and Wildlife Service
Erin Dean, agent in charge, U.S. Fish and Wildlife Service, Los Angeles
Skip Wissinger, special agent (retired), U.S. Park Service

Biologists

Don Rothaus, shellfish biologist, dive team, WDFW
Brent Vadopalas, shellfish biologist, University of Washington
Alex Bradbury, shellfish biologist, WDFW
Lobo Orensanz, shellfish biologist, University of Washington
Daniel Pauly, fisheries biologist, University of British Columbia
Jennifer Jacquet, fisheries biologist, University of British Columbia
Are Strom, shellfish biologist, WDFW
Juan Valero, shellfish biologist, University of Washington
Tim Essington, fisheries biologist, University of Washington
Wayne Palsson, fisheries biologist, WDFW
Bob Sizemore, shellfish biologist, WDFW
C. Lynn Goodwin, shellfish biologist (retired), WDFW
Suzanne Kohin, shark biologist, National Marine Fisheries Service
Vincent Gallucci, ocean and fisheries sciences, University of Washington
Jorge Morales Guiza, fisheries biologist, Baja, Mexico
Glenn VanBlaricom, University of Washington
Pat Gearin, marine mammal biologist, NOAA
Gregor Cailliet, Moss Landing Marine Labs
Chris Lowe, Shark Lab, California State University at Long Beach
Trey Knott, forensic biologist, NOAA Laboratory, Charleston, South Carolina
M. Katherine Moore, biologist, NOAA Research Lab, Charleston
Lara Adams, forensic biologist, NOAA Laboratory, Charleston

Geoduck divers

Gene Canfield, geoduck diver
Casey Bakker, shellfish buyer, diver
Don Webster, geoduck diver, Hodgson's first partner
Carl Sheats, diver, son of Robert Sheats

Mark Mikkelsen, geoduck diver
Steve Sigo, geoduck diver, Squaxin Island Tribe
Craig Parker, geoduck diver
Connie Whitener, geoduck diver
Heidi Mills, geoduck diver
Others preferred to remain unnamed

Attorneys

Micki Brunner, assistant U.S. attorney, Seattle
Maureen Bessette, assistant U.S attorney, Oakland
Joseph Johns, assistant U.S. attorney, Los Angeles
Karolyn Klohe, (former) assistant attorney general, State of Washington
Joe Panesko, assistant attorney general, State of Washington
Marilyn Brenneman, chief, fraud division, King County Prosecutor's Office
Tom Moore, Pierce County deputy prosecutor

Seafood experts

Claude Tchao, CEO Tri-Star Seafood, Vancouver, B.C.
Bob Donegan, president, Ivar's
Quentin Fong, seafood marketing specialist, University of Alaska
Noritoshi Kanai, president, Mutual Trading Company
Seicho Fujikawa, Mutual Trading
Tom Douglas, chef
Charles Ramseyer, chef
Jon Rowley, seafood marketer
Shiro Kashiba, sushi chef
Harry Yoshimura, Mutual Fish Company
Mark Wen, seafood aficionado
Tony Wong, geoduck broker, Vancouver

Others

Doug Tobin
Martin Cetron, global migration and quarantine, Centers for Disease Control
Craig Hoover, U.S. Fish and Wildlife Service
Crawford Allan, North American director, TRAFFIC
Paul Watson, Sea Shepherd Conservation Society

Michael Muehlbauer, wildlife inspector, USFWS
Adrian Lugo, contractor
Thomas Dades, detective (retired), New York Police Department
Nancy Lyles, friend of Tobin's
Jim Peters, Squaxin Island Tribe
Gigi Hogan, whose family once owned Pearls by the Sea
Dennis Lucia, diving instructor
Paul Dorpat, biographer, Ivar Haglund
Dave Hearn, geoduck broker
Todd Palzer, geoduck chief, Department of Natural Resources
Rory Gilliland, former chief, Squaxin Island Tribe
Charles Amner, manager, Mitzels in Fife
Jeffrey Albulet, former executive, Clearbay Fisheries
Roy Ewen, geoduck digger
Jerry McCourt, English instructor
James Bartels, friend of DeCourville and Rick Jones
Edward Binder, artist and Tobin acquaintance
Jerry Elfendahl, coauthor, "The Gooey Duck Song"
Duane Pasco, artist
Linda Gentille, ex-wife, Nichols DeCourville
Darin Carroll, Centers for Disease Control
Jennifer McQuiston, Centers for Disease Control

Acknowledgments

When I set out to write this book I didn't know just how much I would be asking of others—nor how gracious and helpful so many people would be. I am grateful to the editors at the *Seattle Times*. They gave me the time to research and write when such a request was asking a lot and never expressed anything but support. They are a spectacular bunch of people. Laura Helmuth and Sarah Zielinski at *Smithsonian* magazine embraced an early version of the geoduck story and shared my sense of wonder about these unusual clams.

Friend and photographer Tom Reese climbed down from his ladder in the middle of a work day and dug through his Puget Sound archives to find the perfect cover photo. Friend and artist Whitney Stensrud turned my harebrained ideas for illustrating the book into something truly exceptional: Only Whitbob could make geoducks and sturgeon look not just tasteful but classy.

This story would not have seen print if not for my agent, Wendy Strothman. Wendy believed in this idea from the start, and never wavered in her support for it or me. I am incredibly lucky to have found her. Marjorie Braman is the kind of editor every writer seeks. With a simple e-mail nudge or phone call she reminded me what a difference an editor's enthusiasm can make. When Marjorie changed jobs, Henry Ferris at William Morrow stepped in and offered smart suggestions at crucial moments that helped reshape the manuscript and bring it in for a landing—even though his own plate was already plenty full. His talented assistant, Danny Goldstein, carefully reviewed every page and offered wise, creative counsel.

I also must thank a handful of great lawyers. Tom Ward at HarperCollins, and Noelle Kvasnosky and Bruce Johnson at Davis Wright Tremaine helped me out with legal issues and made the process pleasant.

I am fortunate to know many fabulous writers and readers, and I benefited greatly from their willingness to share their talent and insight. For early reads and suggestions on chapters, I thank Ross Anderson, Ken Armstrong, Hal Bernton, Alex Bradbury, Chris Cousins, Sarah Flynn, Deb Gruver, Kristi Heim, Ian Ith, Ian Johnson, Lewis Kamb, Mark Kramer, Jim Langston, Lynda Mapes, Jonathan Martin, Maureen O'Hagan, Dan Niemi, Nick Perry, Casey Seiler, Chris Solomon, Eric Sorensen, and Brent Vadopalas. A handful of friends went far beyond what I had any right to ask. Grace Hobson, Jim Lynch, Brent Shepherd, and John Zebrowski burrowed deep into multiple drafts, talked me through trouble spots again and again, and always sent me back to the keyboard grateful and enthusiastic. Jerry Holloron, copy editor extraordinaire, gave the final manuscript the kind of attention that only he could and did it graciously and on deadline. I owe much to James Scott, who not only read and edited several versions of the manuscript, but also managed with every phone call—and there were dozens—to help me see just what I needed to do next.

The support from friends and family, though less tangible, was equally important. Barb, Bob, Lisa, and Wanda never lost faith and found endless ways large and small to be of help. Their confidence was so consistent and sincere that I was able to draw upon it when my own ran low.

But mostly I am thankful to my wife, Jennifer. She edited every draft, held life together while I stared at the keyboard, and managed it all with her usual warmth and quiet grace. And deep into the endless reporting trips and phone calls and hours of rewriting—when she was busy managing our most important project—she was *still* willing to gamble on me and my ideas. There is no greater gift.